METABOLISM AND REGULATION OF SECONDARY PLANT PRODUCTS

Recent Advances in Phytochemistry
Volume 8

CONTRIBUTORS

LeRoy L. Creasy

Heinz G. Floss

Leslie Fowden

T. Galliard

T. R. Green

H. Grisebach

K. Hahlbrock

Peter F. Heinstein

F. Loewus

A. M. D. Nambudiri

James E. Robbers

C. A. Ryan

H. A. Stafford

G. H. N. Towers

C. P. Vance

Milton Zucker

METABOLISM AND REGULATION OF SECONDARY PLANT PRODUCTS

Recent Advances in Phytochemistry
Volume 8

Edited by

V. C. RUNECKLES

Department of Plant Science
University of British Columbia
Vancouver, British Columbia, Canada

and

E. E. CONN

Department of Biochemistry and Biophysics
University of California at Davis
Davis, California

ACADEMIC PRESS New York San Francisco London *1974*

A Subsidiary of Harcourt Brace Jovanovich, Publishers

ACADEMIC PRESS, INC.
111 Fifth Avenue, New York, New York 10003

United Kingdom Edition published by
ACADEMIC PRESS, INC. (LONDON) LTD.
24/28 Oval Road, London NW1

LIBRARY OF CONGRESS CATALOG CARD NUMBER: 77–182656

ISBN 0–12–612408–6

PRINTED IN THE UNITED STATES OF AMERICA

CONTENTS

Phenylalanine Ammonia-Lyase and Phenolic Metabolism

Leroy L. Creasy and Milton Zucker

Enzymology and Regulation of Flavonoid and Lignin Biosynthesis in Plants and Plant Cell Suspension Cultures

H. Grisebach and K. Hahlbrock

Possible Multienzyme Complexes Regulating the Formation of C_6–C_3 Phenolic Compounds and Lignins in Higher Plants

H. A. Stafford

v

Photoregulation of Phenylpropanoid and Styrylpyrone Biosynthesis in *Polyporus hispidus*

G. H. N. Towers, C. P. Vance, and A. M. D. Nambudiri

Nonprotein Amino Acids from Plants: Distribution, Biosynthesis, and Analog Functions

Leslie Fowden

Proteinase Inhibitors in Natural Plant Protection

C. A. Ryan and T. R. Green

Regulatory Control Mechanisms in Alkaloid Biosynthesis

Heinz G. Floss, James E. Robbers, and Peter F. Heinstein

The Biochemistry of *myo*-Inositol in Plants

F. Loewus

Unusual Fatty Acids in Plants

T. Galliard

LIST OF CONTRIBUTORS

Numbers in parentheses indicate the pages on which the authors' contributions begin.

LEROY L. CREASY (1), Department of Pomology, Cornell University, Ithaca, New York

HEINZ G. FLOSS (141), Department of Medicinal Chemistry and Pharmacognosy, School of Pharmacy and Pharmacal Sciences, Purdue University, West Lafayette, Indiana

LESLIE FOWDEN (95), Rothamsted Experimental Station, Harpenden, Hertfordshire, England

T. GALLIARD (209), Agricultural Research Council, Food Research Institute, Norwich, England

T. R. GREEN (123),* Department of Agricultural Chemistry, Washington State University, Pullman, Washington

H. GRISEBACH (21), Biological Institute II, University of Freiburg i. Br., Germany

K. HAHLBROCK (21), Biological Institute II, University of Freiburg i. Br., Germany

PETER F. HEINSTEIN (141), Department of Medicinal Chemistry and Pharmacognosy, School of Pharmacy and Pharmacal Sciences, Purdue University, West Lafayette, Indiana

F. LOEWUS (179), Department of Biology, State University of New York at Buffalo, Buffalo, New York

A. M. D. NAMBUDIRI (81), Botany Department, University of British Columbia, Vancouver, British Columbia

* Present address: Department of Chemistry, University of California at Los Angeles, Los Angeles, California.

ix

JAMES E. ROBBERS (141), Department of Medicinal Chemistry and Pharmacognosy, School of Pharmacy and Pharmacal Sciences, Purdue University, West Lafayette, Indiana

C. A. RYAN (123), Department of Agricultural Chemistry, Washington State University, Pullman, Washington

H. A. STAFFORD (53), Biology Department, Reed College, Portland, Oregon

G. H. N. TOWERS (81), Botany Department, University of British Columbia, Vancouver, British Columbia

C. P. VANCE (81), Botany Department, University of British Columbia, Vancouver, British Columbia

MILTON ZUCKER (1),* Washington State University, Pullman, Washington

* Deceased.

PREFACE

The thirteenth annual meeting of the Phytochemical Society of North America was held August 8–10, 1973, at Asilomar State Park and Conference Center in Pacific Grove, California. The theme of the annual symposium was "The Metabolism and Regulation of Secondary Plant Products"; the invited papers presented there are collected in this volume.

The chemical history of secondary plant products covers almost the same period of time as the field of organic chemistry itself. Indeed, these unusual plant compounds attracted the interest and efforts of some of the founders of organic chemistry in the first half of the nineteenth century. On the other hand, the biochemical examination, or more specifically, the metabolic study of secondary plant compounds, has not proceeded simultaneously with the rapid development of that subject in animals and microorganisms. While the main features of the metabolism of primary cell metabolites were already established by the 1950's, it is only in the last two decades that the study of secondary plant metabolism has been initiated, and even today much remains to be done at the level of enzymology.

The study of the processes by which cell metabolites are regulated has recently become a field of remarkable activity. To be sure, most of the metabolites examined are primary in nature. It seemed to the organizers of this symposium, however, that enough was known of the regulation of some secondary plant products to warrant a symposium devoted to that topic. We wish to thank the contributors for their cooperation in this endeavor. Their work will serve as models for study in the years ahead.

From the beginning the organizers planned that Milton Zucker should present his classic work on phenylalanine ammonia-lyase (PAL) and its role in the regulation of phenylpropanoids as a keynote paper. The chapter by L. L. Creasy and M. Zucker contains material which Dr. Zucker had intended to present, and we are grateful to Dr. Creasy for his completion of the manuscript and the inclusion of some of the most recent work carried out in Dr. Zucker's laboratory. We were deeply saddened when we learned the reason that Milton Zucker could not come to Asilomar; the tragic news of his death was announced to those attending the symposium. We dedicate this volume to his memory.

A volume of this type would not be possible without the contributors, but there are others who make the publication a reality. Particular thanks are due to Ms. Diane Green, who handled the necessary retyping and collation of manuscripts with flair.

V. C. Runeckles
E. E. Conn

MILTON ZUCKER

1928–1973

Milton Zucker was born in St. Louis, Missouri on August 19, 1928. After receiving an A.B. degree from Washington University in St. Louis, he married Monica Ribstein and shortly thereafter joined Professor Barry Commoner of the same university for graduate study. In 1952 Milton received a Ph.D. in Plant Physiology and moved to Johns Hopkins University as a Fellow of the McCollum-Pratt Institute where he did some of the pioneering work on nitrate reductase with Professor Alvin Nason. Two years later the Zuckers moved to New Haven, Connecticut, where he joined the Department of Plant Pathology of the Connecticut Agricultural Experiment Station. Over the next 17 years Milton worked on phenolic metabolism and its control and the biochemistry of host–pathogen inter-action in plants. He made very significant contributions to these areas. In the fall of 1971 he came to Washington State University as the Chairman of the Department of Agricultural Chemistry and Professor of Biochem-istry, and held these positions until leukemia claimed his life on July 29, 1973. He is survived by his four sons and wife Monica.

Milton was a stimulating colleague, and his gentle, unassuming manner left a lasting impression on those who knew him even for a short time. In warmth of personality and concern for others he went beyond his scientific colleagues. Problems of society such as racial prejudice and war concerned him deeply, and he actively participated in attempts to solve them. While in New Haven, he chose to raise his family in a racially integrated neighbor-hood and took a leading role for many years in solving problems of the community. Milton was a lover of music, literature, art, and nature. While in Pullman he combined camping and canoeing with his favorite hobby of bird watching.

The scientific contributions of Milton Zucker are well known to students of phenolic biosynthesis and enzyme turnover in plants. He served the plant science community as a member of the Editorial Board of *Plant Physiology* for eight years. Milton Zucker, pioneering scientist and warm colleague, will live in our fond memories.

PHENYLALANINE AMMONIA-LYASE AND
PHENOLIC METABOLISM*

LEROY L. CREASY

Department of Pomology, Cornell University, Ithaca, New York

MILTON ZUCKER†

Washington State University, Pullman, Washington

Introduction

The synthesis of phenolics in plants results from several different pathways (see sections in Grisebach, 1972). The formation of phenolic rings from acetate (and malonate) by cyclization of polyketide chains accounts for a large number of phenols itself and contributes a significant source of carbon for the production of other compounds, notably the flavonoids. The

---

* Scientific paper no. 4264. Project 0125 from the College of Agriculture Research Center, Washington State University, Pullman, Washington.
  † Deceased.

1

phenylpropanoid compounds derived from the shikimic acid pathway not only occur widely in plants, but are precursors for additional classes of phenolic materials (Freudenberg and Neish, 1968).

One of the more intriguing aspects of the metabolism of phenolic compounds is the rapid changes which can take place in their rates of synthesis. Evidence of the occurrence of specific phenolics only in some limited parts of plants, or only during specific times of the year, or only in specific cell types, or only in response to certain stimuli implies considerable regulation over their synthesis and further metabolism. Thus, the absence of a phenolic in a particular plant material would not suggest that the plant was incapable of synthesizing it. The profound capability of plants to manufacture phenolics in response to the stimuli used by investigators to study their synthesis suggests that synthesis must be subject to strict control under most conditions.

Initially, phenolic research followed the usual pattern of identification of the wide variety of phenolic compounds which are synthesized by plants. This continuing effort has provided an initial look at a valuable plant property (Swain, 1966; Harborne, 1970) and each new technique contributes to further knowledge of plant chemistry. As knowledge of these compounds mounted, the second stage of research occurred, namely, that of deciphering the biosynthetic sequences by which plants manufacture their varied complements of phenolics (Harborne, 1964). A third stage of research concerns the regulation of the metabolism of phenolics. This area has lagged behind the others, certainly in part because the former information is vital in any regulatory studies. The following discussion of regulation will be concerned only with those phenolics which can be derived from phenylalanine and tyrosine through the ammonia-lyases.

The biosynthesis of phenylpropanoids from phenylalanine and tyrosine is initiated through the action of ammonia-lyases on phenylalanine (PAL) (Koukol and Conn, 1961) and on tyrosine (TAL) (Neish, 1961). Most of the research has been done on PAL because it is found more frequently (Camm and Towers, 1973a). Many aspects of the enzymology of PAL (Hanson and Havir, 1972) and the physiology of its appearance in plants have been studied. Several reviews are available which cover recent literature on phenylalanine ammonia-lyase (Zucker, 1972; Camm and Towers, 1973a). It is our intent to discuss only those areas related to control mechanisms of phenolic synthesis.

Rates of Phenolic Biosynthesis

The importance of phenolic compounds in the overall carbon metabolism of plants can be appreciated by considering their rates of synthesis. Proposed pathways of phenolic biosynthesis or regulatory schemes thereof can be evaluated by correlating *in vitro* enzyme activity with *in vivo* rates of synthesis of the appropriate phenolics (Ahmed and Swain, 1970; Swain and Williams, 1970; Creasy, 1968a). Thus in the potato tuber, where chlorogenic acid is the major phenolic (Hanson and Zucker, 1963), a turnover rate of 50 nmoles/hr per gram fresh weight was observed with tissues having a relatively fixed content. At a stable concentration of chlorogenic acid in *Xanthium* leaf disks the turnover was 100 nmoles/hr per gram fresh weight (Taylor and Zucker, 1966). These rates of turnover agree well with the maximum initial rates of chlorogenic acid synthesis observed in potato tuber disks (Zucker, 1963, 1965). The maximum rate of synthesis of anthocyanins in *Sorghum* seedlings was 28 nmoles/hr per gram fresh weight, although other phenolics were not measured (Stafford, 1966), and the turnover of anthocyanins may have made this rate a low estimate (Steiner, 1972). Excised pods of *Phaseolus vulgaris* synthesized phaseollin at 26 nmoles/hr per gram in response to inducers of this phytoalexin (Hess and Hadwiger, 1971). In strawberry leaf disks the accumulation of cinnamic acids, flavonoids, and tannins was determined to be equivalent to the utilization of 2.4 μmoles of C_6–C_3 units per hour per gram fresh weight (Creasy, 1968a). This rate of utilization did not include any turnover and is a minimal estimate. It was observed that the rate of carbon utilized approximated that of respiration. In similar experiments in which the utilization of carbohydrate was followed, a large proportion of that lost was committed to the synthesis of phenolics (Creasy, 1968b). Ahmed and Swain (1970) reported an accumulation rate of 1.68 μmoles phenylpropanoids/hr per gram fresh weight in illuminated mung bean seedlings. These data clearly show that the synthesis of phenolics can constitute a major part of a plant's metabolism, but leave unanswered the question of what, if any, advantage is conferred upon the plant by such metabolism.

Substrate Control

Phenylalanine

Phenylalanine, a ubiquitous aromatic amino acid, and its companion, tyrosine, serve as the initial substrates for many plant phenolics. Although other pathways to phenolic compounds exist [e.g., acetate(malonate)-derived phenolic rings (Grisebach, 1972)] and other sequences for their synthesis are possible (Swain and Williams, 1970), phenylalanine and tyrosine are utilized for the synthesis of the polypeptides, phenolic acids, flavonoids, lignin, etc. The first step involving phenylalanine in most of these biosyntheses is catalyzed by phenylalanine ammonia-lyase. The extent to which tyrosine ammonia-lyase is responsible for deaminating tyrosine in phenolic biosynthesis remains unanswered. It has been proposed that one ammonia-lyase is involved and this is supported in studies of maize (Havir *et al.*, 1971) and wheat (Nari *et al.*, 1972) enzymes.

The shikimic acid pathway results in the production not only of phenylalanine and tyrosine but also the amino acid tryptophan, which, in addition to being a protein amino acid, is a precursor of a number of secondary plant products. The regulation of the synthesis of these three amino acids and their subsequent utilization must successfully cope with different demands for numerous products at different stages in growth and differentiation.

There is a paucity of data on the content of phenylalanine in tissues during the onset of rapid phenolic biosynthesis. Studies by Amrhein and Zenk (1971) showed a remarkably constant level of both phenylalanine and tyrosine in tissues under conditions where the rapid synthesis of phenolics was taking place. One of the present authors (L. L. Creasy, unpublished data) has found the content of phenylalanine to be constant in leaf disks which were treated to cause the accumulation of cinnamic acid (Creasy, 1971). In germinating pumpkin seedlings a constant level of phenylalanine was observed (Chou and Splittstoesser, 1972). With large changes in the rate of utilization of phenylalanine required for phenolic synthesis, very sensitive controls of its level in plants must exist. This in turn suggests that the rate of production and not the concentration of phenylalanine is the controlling factor in phenolic biosynthesis. The regulation of phenylalanine production may differ between microorganisms, where its metabolism was first studied, and higher plants. Feedback inhibition is the primary regulatory mechanism in microorganisms (Cotton and Gibson, 1965; Gibson and Pittard, 1968). Multifunctional proteins are known, such as the T-protein and P-protein of *Aerobacter aerogenes* (Cotton and Gibson, 1968b; Koch *et al.*, 1972), which catalyze the synthesis of tyrosine and phenylalanine, respectively, from chorismic acid and are individually regulated by feedback of the appropriate amino acid. Additional allosteric effects (Monod *et al.*, 1963) have been reported in the prephenate dehydratase of *Bacillus subtilis* which exhibits feedback inhibition by phenylalanine. Allosteric specificities are exhibited for methionine (activator), leucine (activator), tryptophan (inhibitor), and tyrosine (reversal of tryptophan inhibition) (Rebello and Jensen, 1970). Such complex metabolic interdependence is characteristic of the regulation of aromatic amino acids in microorganisms and confirms that careful studies of their control mechanisms are justified in higher plants (Gilchrist *et al.*, 1972).

In higher plants the biosynthetic pathways of aromatic amino acids are

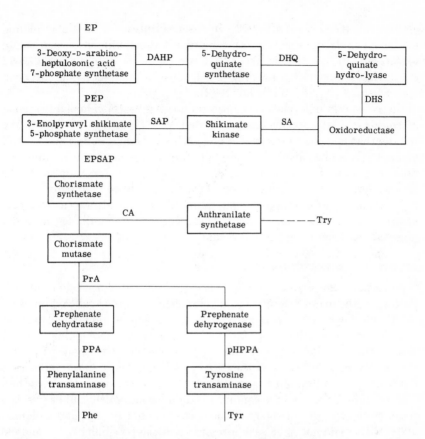

FIG. 1. Synthesis of aromatic amino acids. Abbreviations: EP, erythrose 4-phosphate; PEP, phosphoenolpyruvate; DAHP, 3-deoxy-D-arabino-heptulosonic acid 7-phosphate; DHQ, 5-dehydroquinic acid; DHS, 5-dehydroshikimic acid; SA, shikimic acid; SAP, shikimic acid 5-phosphate; EPSAP, 3-enolpyruvyl shikimic acid acid 5-phosphate; CA, chorismic acid; PrA, prephenic acid; PPA, phenylpyruvic acid; pHPPA, *p*-hydroxy-phenylpyruvic acid; Phe, phenylalanine; Tyr, tyrosine; Try, tryptophan.

essentially the same as in microorganisms (Gamborg, 1966; Gibson and Pittard, 1968) (Fig. 1). The chorismate mutase from pea seedlings was reported to be inhibited by phenylalanine and tyrosine and activated by tryptophan (Cotton and Gibson, 1968a). Two isozymes of chorismate mutase were found in etiolated mung bean; one isozyme was inhibited by both phenylalanine and tyrosine, and the inhibition reversed by trypto-phan (Woodin and Kosuge, 1968; Gilchrist *et al.*, 1972). In this regard it resembles the enzyme from green algae (Weber and Böck, 1969) and *Neurospora* (Baker, 1966). The second isozyme was unaffected by the three amino acids (Gilchrist *et al.*, 1972). In tissue cultures of three higher plants, neither phenylalanine nor tyrosine repressed the synthesis of chorismate mutase and therefore it is likely that regulation of aromatic amino acid synthesis is affected principally by feedback and metabolic interlock mechanisms (Chu and Widholm, 1972).

The constancy of levels of aromatic amino acids reported in plants is good reason to suspect effective control mechanisms. However, it is interesting to note that no one has reported a phenolic compound to exert any control of aromatic amino acid synthesis, although these amino acids must be rapidly produced. The aromatic amino acids, therefore, appear to be independent of the regulation of phenolic metabolism and vice versa. Subsequent transformations of the aromatic amino acids appear to represent a distinct separation in metabolism suggestive of a phylogenetically unique biosynthetic sequence (Fraenkel, 1959).

SUBSTRATE "INTERACTION"

Some phenolic products are esterified to products of other biosynthetic pathways, the major example being chlorogenic acid (3-*O*-caffeoylquinate). What role might the regulation of the production of quinic acid play in controlling the synthesis of chlorogenic acid or dicaffeoylquinic acid or caffeoylglucose? Plants overloaded with phenolic acids frequently synthesize the glucose esters of those phenols. Can this be interpreted as a detoxification rather than a normal sequence of biosynthesis (Harborne and Corner, 1961)? Some plants contain free quinic acid which could serve as a reserve for synthesis of esters (Boudet *et al.*, 1967; Boudet and Colonna, 1968). Others do not have such a reserve and the possibility of some cross pathway control system seems likely. The physiological role of quinic esters is of concern since they apparently do not serve directly as precursors of subsequent metabolites (Alibert *et al.*, 1972b; Majak and Towers, 1973).

Enzyme Regulation

SHIKIMIC ACID PATHWAY

The shikimic acid pathway has been well established in higher plants both by labeling studies with radioisotopes and by the demonstration of the necessary enzymes. The control of the enzymes of this pathway is now being studied. A multienzyme complex has been reported for *Quercus* roots (Boudet, 1971) which has the activity of 5-dehydroquinate hydro-lyase and shikimate:NADP oxidoreductase (Fig. 1).

Some enzymes of the shikimic acid pathway are reported to increase under conditions which lead to phenol production in sweet potato root slices (Kojima *et al.*, 1969; Minamikawa *et al.*, 1966a,b), potato tuber slices

(Camm and Towers, 1973a), and mung bean seedlings (Ahmed and Swain, 1970). The specific activity of 5-dehydroshikimic reductase in tea plants was greater in the actively growing shoot tip than in mature leaves (Sanderson, 1966). Conversely, the activity of shikimic acid:NADP oxidoreductase was not changed by light in the seedlings of species studied by Amrhein and Zenk (1971), although they examined some species that were known to synthesize phenolics rapidly on exposure to light. Pea seedlings have also shown stable levels of this enzyme during light treatment, resulting in phenolic synthesis (Attridge and Smith, 1967; Ahmed and Swain, 1970). The 3-deoxy-D-arabino-heptulosonic acid-7-phosphate synthetase of sweet potato root remained constant after slicing (Minamikawa and Uritani, 1967). The changes in enzymes reported in some tissues are possibly a consequence of renewed metabolic activity resulting in the revitalization of many synthetic pathways and are not directly related to transient stimulation of phenolic biosynthesis. It, therefore, appears possible that in many cases of induced phenolic biosynthesis the shikimic acid pathway enzymes do not change and no direct control connections exist between them and phenolic biosynthesis.

Phenylalanine Ammonia-Lyase

Phenylalanine ammonia-lyase has received most of the attention in studies of the regulation of phenolic biosynthesis. In general the activity of PAL increases dramatically when plants are subjected to conditions resulting in the stimulation of phenolic synthesis (see references in Camm and Towers, 1973a). The question of whether the quantity of PAL is the limiting factor in total phenolic synthesis is not clearly resolved. It has been reported that the addition of phenylalanine will stimulate chlorogenic acid synthesis while suppressing the increase in PAL activity in potato tubers (Zucker, 1965). Exogenously supplied phenylalanine also stimulated the accumulation of phenolic acids in strawberry leaf disks without stimulating PAL activity (Creasy, 1971). If, however, the concentration of phenylalanine is normally rigidly maintained in a tissue (Amrhein and Zenk, 1971), then increases in PAL activity would still cause increased utilization of phenylalanine for phenolic biosynthesis, even though phenylalanine were not at an optimum concentration. In this instance phenylalanine might be limiting but PAL activity could still be regulating the total flow of carbon into phenolics. Regulation of PAL activity would appear to be a likely metabolic point for control.

Induction and Repression of de Novo Synthesis

The single requirement for deciding whether or not PAL is an induced enzyme is that it be synthesized *de novo*. The increase in PAL activity

initiated by light was shown by the deuterium bouyant density technique
to be *de novo* synthesis in potato tuber (Sacher *et al.*, 1972), gherkin seedlings
(Iredale and Smith, 1973), and mustard seedlings (Schopfer and Hock,
1971). In *Xanthium* leaves, PAL was found to be radioactive after feeding
labeled amino acids (Zucker, 1969). Other claims of *de novo* synthesis are
based on the response to various inhibitors (Engelsma, 1967; Minimikawa
and Uritani, 1965; Scherf and Zenk, 1967; Walton and Sondheimer, 1968;
Zucker, 1965). In bean pod tissue, PAL activity can be increased by a
number of DNA-intercalating agents (Hess and Hadwiger, 1971), suggesting a requirement for newly synthesized RNA. The level of PAL in tissues
may depend on a number of other factors in addition to the rate of induced
synthesis (e.g., inactivation) and only in a limited number of plants might
induction be a regulatory mechanism. The identity of the inducer or
inducers of PAL is of considerable interest due to the wide variety of
stimuli which result in increases in PAL level (Camm and Towers, 1973a).

There is significantly less known about the repression of induced synthesis than of the occurrence of *de novo* synthesis. Although it has been
reported that phenolic acids can repress PAL synthesis (Engelsma, 1968)
and may be related to photoinduction (Engelsma, 1972), it was also observed that they increased inactivation (Engelsma, 1968). The inactivation
of PAL was shown not to affect the synthesis of PAL in mustard seedlings
(Weidner *et al.*, 1969). The existence of end product repression of PAL is
therefore still an open question and understandably so, considering the
differences in accumulated end products in different plant tissues. A more
general mechanism for PAL regulation must exist.

The increase in PAL activity resulting from light in some tissues has
been proposed to be due to activation of previously formed inactive PAL
protein. The cycloheximide-stimulated PAL increase in *Cucumis* seedlings
(Attridge and Smith, 1973) was interpretated as arising from a pool of
inactive PAL which was synthesized during or after imbibition (Iredale
and Smith, 1973). In radish cotyledons an inactive PAL was found to be
synthesized in the dark and activated on exposure of the cotyledons to
far-red light (Klein-Eude *et al.*, 1974) and the inactive PAL protein was
isolated from dark-grown radish by techniques of protein chemistry
(Blondel *et al.*, 1973).

PAL Turnover

The concept of turnover of enzymes as a mechanism of regulation is well
established for a number of animal enzyme systems (Schimke, 1969). In
one enzyme, tryptophan oxygenase, application of a hormone results in
an increased enzyme level due to an increased rate of synthesis. In addition,

the administration of tryptophan leads to an increased enzyme level as a result of decreased inactivation (Schimke *et al.*, 1965). An inactivation half-life of 2 hours assures rapid regulation of tryptophan oxygenase even with high rates of synthesis. Enzyme turnover was considered to be a regulation phenomenon involving organelles (Dehlinger and Schimke, 1971; Dice and Schimke, 1972), and the inactivation was thought to be due to proteolysis with an increased rate of degradation associated with larger proteins (Dehlinger and Schimke, 1970). Specific proteases for pyridoxal enzymes (Katunuma *et al.*, 1971a) and for NAD-dependent dehydrogenases (Katunuma *et al.*, 1971b) have been reported and the inactivation in each case is prevented by the appropriate cofactors. A different mechanism of inactivation occurs in the regulation of invertase. Changes in enzyme activity are brought about by reversible coupling of a specific protein with the enzyme (Pressey and Shaw, 1966; Pressey, 1967).

The inactivation of PAL initially reported in potato tuber disks (Zucker, 1965) has been observed in a number of different plant materials: *Xanthium* (Zucker, 1969), gherkin seedlings (Engelsma, 1967, 1968; Attridge and Smith, 1973), citrus fruits (Riov *et al.*, 1969), buckwheat seedlings (Scherf and Zenk, 1967; Amrhein and Zenk, 1970), tulip anthers (Wiermann, 1973), mustard seedlings (Weidner *et al.*, 1969), excised bean axes (Walton and Sondheimer, 1968), and sunflower leaf disks (L. L. Creasy, unpublished data). Unlike most of the above examples of inactivation, PAL loss in asparagus spears (Goldstein *et al.*, 1972) and in *Oenothera* seedlings (Hachtel and Schwemmle, 1972) was not inhibited by cycloheximide.

These observations have led to several hypotheses on the mechanism of inactivation and its role in the regulation of PAL activity. Engelsma (1970) has proposed a concept of a PAL–inhibitor complex similar to the invertase system (Pressey, 1967). He proposes that PAL is inactivated by reacting with an inducible proteinaceous inhibitor which under conditions of cold treatment is caused to dissociate when the temperature is raised to normal (Engelsma, 1969). This hypothesis is consistent with his observations that (1) PAL synthesis and inactivation are inhibited by cycloheximide (Engelsma, 1967), (2) the reappearance of PAL following light induction, decay, and cold treatment is cycloheximide-insensitive (Engelsma, 1969), and (3) an inhibitor of PAL is found in gherkin seedlings (Engelsma and van Bruggen, 1971). This hypothesis has been expanded by Attridge and Smith (1973), who invoke a pool of inactive PAL, although its synthesis could not be determined unequivocally by bouyant density labeling (Iredale and Smith, 1973). The possibility of the interconversion of active and inactive forms of PAL has been proposed by Havir and Hanson (1968, 1973).

An alternative hypothesis based on proteolytic degradation of PAL has been advanced. Zucker (1969) noted the loss of radioactivity from labeled PAL during inactivation and suggested that the protein molecule was being degraded. Further studies in *Xanthium* showed that synthesis of the PAL protein, as measured by the incorporation of labeled amino acids, occurred both in light during a net increase in activity and in darkness when the amount of activity was decreasing (Zucker, 1970, 1971). The influence of light, therefore, was on the rate of inactivation rather than the rate of synthesis. We have recently been examining the turnover of PAL in *Helianthus annus* leaf disks (L. L. Creasy, unpublished data). These leaf disks are light-insensitive if supplied with sucrose and show a sucrose-dependent increase in PAL in the dark. When sucrose is removed there is a rapid, first-order loss of PAL activity with a half-life of 2 hours. The PAL can be increased again at any time in leaf disks by the addition of sucrose (Fig. 2). Both the increase and the loss of PAL can be prevented by 100

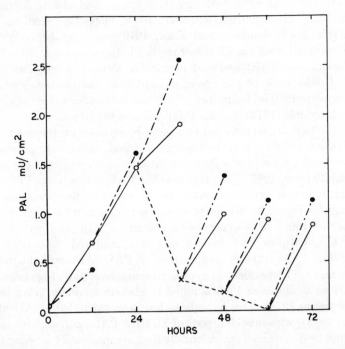

FIG. 2. Effect of cycloheximide on the activity of phenylalanine ammonia-lyase in *Helianthus annus* leaf disks. Disks were transferred to 10 µg/ml cycloheximide in 0.1 M sucrose (●) or to sucrose solutions (○) every 12 hours from either sucrose or from water (×). (From L. L. Creasy, M. Zucker, and P. P. Wong. 1974. *Phytochemistry* 7. Courtesy of Pergamon Press, Ltd.)

μg/ml of cycloheximide. The use of lower cycloheximide concentrations revealed that the loss of PAL was more sensitive to cycloheximide than the increase in activity, and 10 μg/ml inhibited the initial increase in activity only 50 percent but inhibited the subsequent loss 90 percent. The inhibition of protein synthesis by cycloheximide as measured by the incorporation of labeled isoleucine was constant throughout the time intervals used in these experiments.

An anomalous effect of cycloheximide was found when it was added at different times in the sequence of PAL increase and loss. The initial inhibition of synthesis by cycloheximide changed to a stimulation in the appearance of PAL (Fig. 2) in sucrose-treated leaf disks. Although cycloheximide inhibited the loss of PAL on water, when disks were returned to sucrose solutions 10 μg/ml of cycloheximide was still able to stimulate the appearance of PAL, while 100 μg/ml of cycloheximide still fully inhibited the appearance. These results suggest that the regulation of PAL activity is due to adjustments in the rate of inactivation due to a loss of the PAL protein. Initially when PAL is "induced" there is little inactivation system present. This was confirmed by the very slow loss of activity from fresh disks ($t_{1/2} > 1$ day). As the level of PAL increased, an inactivating system was formed which finally limited further net increases in the activity of PAL. Application of cycloheximide resulted in a stimulation of PAL activity because of its differential inhibition of synthesis and inactivation. The observed rate of PAL increase in sucrose with 10 μg/ml of cycloheximide would be 50 percent of the total rate of PAL formation if the effect of cycloheximide were the same as on initial synthesis. The total rate calculated this way was the same as the total rate calculated by correcting the net rate of increase in sucrose for the initial rate of loss on transfer to water. If there had been a release of PAL from an inactive PAL–inactivator complex due to cycloheximide treatment, we would have found an increase in activity during PAL inactivation on water when cycloheximide was added. This did not occur and cycloheximide treatment during water inactivation resulted in the 90 percent inhibition of inactivation as expected (Fig. 3).

It was also found that L-phenylalanine mimicked the response of cycloheximide to some degree. It had been observed that phenylalanine applied during induction resulted in the inhibition of PAL appearance. We found, however, that when phenylalanine was applied later in the sequence we observed a stimulation in the rate of appearance of PAL on sucrose and an inhibition of the loss of PAL when sucrose was removed (Table 1). The action of phenylalanine in protecting PAL was confirmed in experiments on the *in vitro* inactivation of PAL (Fig. 4) in which it was also shown that PAL could be separated from a heat-labile inactivator by ammonium sulfate fractionation (L. L. Creasy, unpublished data) (Table 2).

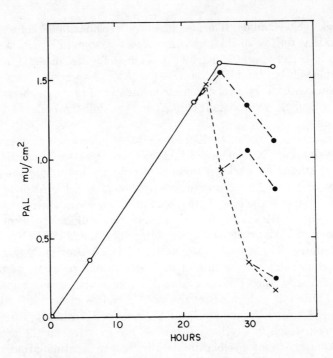

Fɪɢ. 3. The effect of 10 μg/ml of cycloheximide in water supplied during the inactiva-tion of phenylalanine ammonia-lyase in *Helianthus annus* leaf disks. After pretreatment with 0.1 *M* sucrose (○), the leaf disks were transferred to water (×) and then at inter-vals during inactivation to cycloheximide in water (●). (From L. L. Creasy, M. Zucker, and P. P. Wong. 1974. *Phytochemistry* **13**. Courtesy of Pergamon Press, Ltd.)

The concept of protection of PAL from inactivation by phenylalanine suggests a complete hypothesis for the regulation of PAL in plant tissues. Induction of PAL by various stimuli results in an increase in its rate of synthesis or activation which is followed by an increase in its inactivation system. It is possible that this increase in inactivation is regulated by products of PAL (Engelsma, 1968) or by some other mechanism such as depletion of a pool of phenylalanine. The inactivation rate increases both because of a net increase in its activity and because of increased PAL content since it is a first-order reaction. A point is reached where synthesis is balanced by inactivation and no net change in activity occurs until synthesis stops (removal of stimuli, e.g., sucrose in *Helianthus*) or inactiva-tion is prevented (increase in protector, e.g., phenylalanine in *Helianthus*). This imbalance causes a new steady-state level which is maintained until some further metabolic event changes the ratio of PAL synthesis and inactivation. The net result is a system in which rapid changes can occur in

TABLE 1

EFFECT OF ADDING PHENYLALANINE AND CINNAMIC ACID TO LEAF DISKS AT DIFFERENT STAGES IN THE *In Vivo* APPEARANCE AND INACTIVATION OF PHENYLALANINE AMMONIA-LYASE[a]

Pretreatment	PAL activity after pretreatment (mU/cm²)	Treatment	Time (hr)	PAL activity after addition (mU/cm²)		
				None	10 mM Phenylalanine	10 mM Cinnamic acid
None	0.131	Sucrose	20	0.819	—	0.556
None	0.185	Sucrose	20	2.095	1.422	—
24 hr sucrose	2.095	Sucrose	6	2.211	3.092	—
24 hr sucrose	2.095	Water	6	0.201	1.832	—
17 hr sucrose	0.989	Water	12	0.464	1.453	—
22 hr sucrose	1.724	Water	12	0.077	0.758	0.557

[a] The PAL activity was measured by extracting the disks at the end of the indicated times. All sucrose treatments were 0.1 M.

TABLE 2

AMMONIUM SULFATE FRACTIONATION OF PHENYLALANINE AMMONIA-LYASE (PAL) AND PHENYLALANINE AMMONIA-LYASE INACTIVATING SYSTEM (PAL-IS)[a]

Disk pretreatment	Ammonium sulfate fraction	PAL activity (mU/mg protein)	PAL-IS activity (k_d)	Inhibition by 25 mM phenylalanine (%)
24 hr sucrose	0–0.2	0.13	—	—
	0.2–0.4	1.80	—	—
	0.4–0.6	0.70	—	—
	0.6–0.8	0	—	—
24 hr sucrose, then 28 hr water	0–0.2	—	0	—
	0.2–0.4	—	0	—
	0.4–0.8	—	0.034	91
	Boiled 0.4–0.8	—	0	—

[a] Sucrose treatments were 0.1 M. Ammonium sulfate fractions are relative to saturation. k_d was determined at 40°C with equal proportions of the starting material using PAL collected from the 0.2–0.4 ammonium sulfate fractionation.

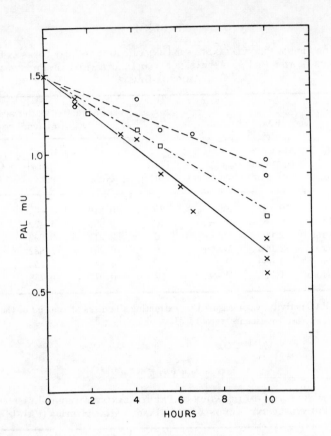

Fig. 4. The *in vitro* loss of phenylalanine ammonia-lyase activity in buffer alone (×) or with 1 m*M* phenylalanine (□) or 10 m*M* phenylalanine (○).

the level of PAL in response to many external stimuli. This hypothesis can also be used to explain the changes in PAL activity of gherkin seedlings (Engelsma, 1970; Attridge and Smith, 1973).

PAL Isozymes

Several investigators have reported the existence of PAL isozymes and related them to the synthesis of different classes of phenolics. Two PAL isozymes were found in sweet potato roots; both were inhibited by cinnamic acid but only one of them was inhibited by *p*-coumaric and caffeic acids (Minimikawa and Uritani, 1965). In mung bean seedlings, separation of PAL by ammonium sulfate resulted in the observation that the light-stimulated PAL was found in only one of the fractions (Ahmed and Swain,

1970). Recent work has described two isozymes of PAL in *Quercus* which differ in their allosteric properties (Boudet *et al.*, 1971; Alibert *et al.*, 1972a). One is principally inhibited by cinnamic acids, while the other is inhibited by benzoic acids; the inhibition in both is reduced by gallic acid. In *Quercus* roots the PAL isozymes were found in different organelles each associated with enzymes leading either to the synthesis of cinnamic acids or to the synthesis of benzoic acids (Alibert *et al.*, 1972a). Considerable particulate PAL activity had been found in sorghum (Stafford, 1969) and the complete conversion of phenylalanine to caffeic acid in a particulate fraction has been shown by Alibert *et al.* (1972b). These findings will stimulate renewed interest in particulate PAL activity even though only a small proportion of total PAL activity is usually associated with particulate fractions (Camm and Towers, 1973b; Amrhein and Zenk, 1971). The relationship of these findings to the concept of multienzyme complexes is discussed by Stafford in this volume.

Cinnamic Acid 4-Hydroxylase

Regulation of phenolic biosynthesis at the level of the hydroxylation of cinnamic acid (Russell and Conn, 1967) has been proposed (Russell, 1971). It has also been reported that the activity of CH is stimulated by light in some higher plants (Russell, 1971; Amrhein and Zenk, 1968; Hahlbrock *et al.*, 1971) and in *Polyporus* (Nambudiri *et al.*, 1973). Increases in activity also result from wounding (Amrhein and Zenk, 1968; Camm and Towers, 1973c) and ethylene treatment (Hyodo and Yang, 1971). In strawberry leaf disks it was shown that the *in vivo* activity of CH greatly exceeded that of PAL and no evidence could be found for a control point (Creasy, 1971). Cinnamic hydroxylase (CH) induced by light and by wounding was of considerably lower activity than the PAL activity of buckwheat seedlings (Amrhein and Zenk, 1968).

In contrast to PAL, cinnamic hydroxylase activity is found primarily in a particulate fraction. Some particulate PAL activity has been reported and it now has been shown that in *Quercus* roots the particles containing CH were those containing one of the PAL isozymes of that tissue (Alibert *et al.*, 1972a,b). However, the particle-bound PAL activity of buckwheat seedlings was not associated with the same particles as the CH activity (Amrhein and Zenk, 1971). In light-treated potato tuber disks, PAL and CH found in a particulate fraction were of similar activities (Camm and Towers, 1973b) and although no measurements of phenolic synthesis were reported, the particulate rates are sufficient to account for rates of phenolic synthesis found by other workers (Zucker and Levy, 1959). In fact, Alibert *et al.* (1972b) have shown the conversion of phenylalanine to caffeic acid by microsomal particles of *Quercus* roots. The concept of bound PAL and CH being a regulatory enzyme complex is, therefore, strengthened (Camm and Towers, 1973b). CH undergoes parallel increases with other enzymes of phenylpropanoid synthesis in parsley cell cultures (Hahlbrock *et al.*, 1971), and it has been suggested that these enzymes are the result of a single operon (Zucker, 1972).

FLAVONOID ENZYMES

The regulation of flavonoid biosynthesis has been extensively examined by Grisebach and Hahlbrock studying flavone synthesis in parsley cell cultures (Hahlbrock *et al.*, 1971), and their work is reviewed by Grisebach in this volume.

Recent research on flavonoid biosynthesis in developing pollen (Wiermann, 1973), isolated protoplasts of *Nemesia* (Hess and Endress, 1973), petals of petunia (Steiner, 1972), callus cultures of carrot (Schmitz and Seitz, 1972), buckwheat seedlings (Amrhein and Zenk, 1971; Laanest and Margna, 1972), and other plant materials should soon provide additional information on control mechanisms of flavonoid synthesis. Flavonoid biosynthesis requires the coordinated control of two major pathways and will undoubtedly be subject to a more complex series of controls than those of the simple phenylpropanoid pathway which we have described here. Enzymes of lignin biosynthesis (Freudenberg and Neish, 1968) are also being characterized which will aid in the description of regulation of lignin biosynthesis (Ebel and Grisebach, 1973; Mansell and Zenk, 1972). It will be interesting to see what control these branches of phenolic metabolism exert on the regulation of phenylalanine ammonia-lyase.

Summary

We have outlined the position of phenylalanine ammonia-lyase in the synthesis of some phenolics in higher plants. It is evident from this outline that a final description of the mechanism of regulation of all phenolic biosynthesis is not yet possible. The lack of information on possible control of the synthesis of aromatic amino acids by phenolics or by stimuli which frequently result in phenolic synthesis indicates that most of the regulation of phenolic synthesis occurs no earlier than at the level of phenylalanine. This is an interesting concept because it effectively isolates phenolic biosynthesis from primary metabolism biosynthetically. The rapid rates of phenolic biosynthesis known to occur in plants must tax the regulatory mechanisms of the aromatic amino acids.

Regulation of phenolic biosynthesis is clearly possible at the level of phenylalanine ammonia-lyase. Present knowledge of the regulation of the concentration of phenylalanine in tissues synthesizing phenolics does not permit the conclusion that phenylalanine itself is normally involved in phenolic regulation. If the levels of phenylalanine are indeed stable following stimulation of phenolic biosynthesis, then the observed changes in PAL activity must be important in allowing for increased phenolic synthesis. We have outlined a possible mechanism to account for rapid regulation of the level of PAL which utilizes the principle of enzyme turnover. This hypothesis is applicable to nongrowing tissues which show transient changes in their rates of phenolic accumulation in response to specific stimuli. While the hypothesis is concerned with the regulation of PAL, it could be applied to other enzymes in phenolic synthesis. It is also clear that at each branch point in the synthesis of the many specific phenolics of plants additional controls must exist, and it is, therefore, of interest to know how these controls interact at the initial step in synthesis.

REFERENCES

Ahmed, S. I., and T. Swain. 1970. *Phytochemistry* **9**:2287–2290.
Alibert, G., R. Ranjeva, and A. Boudet. 1972a. *Biochim. Biophys. Acta* **279**:282–289.
Alibert, G., R. Ranjeva, and A. Boudet. 1972b. *Physiol. Plant.* **27**:240–243.
Amrhein, M., and M. H. Zenk. 1968. *Naturwissenschaften* **55**:394.
Amrhein, N., and M. H. Zenk. 1970. *Z. Pflanzenphysiol.* **63**:384–388.
Amrhein, N., and M. H. Zenk. 1971. *Z. Pflanzenphysiol.* **64**:145–168.
Attridge, T. H., and H. Smith. 1967. *Biochim. Biophys. Acta* **148**:805–807.
Attridge, T. H., and H. Smith. 1973. *Phytochemistry* **12**:1569–1574.
Baker, T. I. 1966. *Biochemistry* **5**:2654–2657.
Blondel, J. D., C. Huault, L. Faye, P. Rollin, and P. Cohen. 1973. *FEBS Lett.* **36**:239–244.
Boudet, A. 1971. *FEBS Lett.* **14**:257–261.
Boudet, A., and J. P. Colonna. 1968. *C. R. Acad. Sci.* **266**:2256–2259.
Boudet, A., G. Marigo, and G. Alibert. 1967. *C. R. Acad. Sci.* **265**:209–212.
Boudet, A., R. Ranjeva, and P. Gadal. 1971. *Phytochemistry* **10**:997–1005.
Camm, E. L., and G. H. N. Towers. 1973a. *Phytochemistry* **12**:961–973.
Camm, E. L., and G. H. N. Towers. 1973b. *Phytochemistry* **12**:1575–1580.
Camm, E. L., and G. H. N. Towers. 1973c. *Can. J. Bot.* **51**:824–825.
Chou, K. H., and Splittstoesser, W. E. 1972. *Physiol. Plant.* **26**:110–114.
Chu, M., and J. M. Widholm. 1972. *Physiol. Plant.* **26**:24–28.
Cotton, R. G. H., and F. Gibson. 1965. *Biochim. Biophys. Acta* **100**:76–88.
Cotton, R. G. H., and F. Gibson. 1968a. *Biochim. Biophys. Acta* **156**:187–189.
Cotton, R. G. H., and F. Gibson. 1968b. *Biochim. Biophys. Acta* **160**:188–195.
Creasy, L. L. 1968a. *Phytochemistry* **7**:441–446.
Creasy, L. L. 1968b. *Phytochemistry* **7**:1743–1749.
Creasy, L. L. 1971. *Phytochemistry* **10**:2705–2711.
Dehlinger, P. J., and R. T. Schimke. 1970. *Biochem. Biophys. Res. Commun.* **40**:1473–1480.

Dehlinger, P. J., and R. T. Schimke. 1971. *J. Biol. Chem.* **246**:2574–2583.

Dice, J. F., and R. T. Schimke. 1972. *J. Biol. Chem.* **247**:98–111.

Ebel, J., and H. Grisebach. 1973. *FEBS Lett.* **30**:141.

Engelsma, G. 1967. *Naturwissenschaften* **54**:319–320.

Engelsma, G. 1968. *Planta* **82**:355–368.

Engelsma, G. 1969. *Naturwissenschaften* **11**:563–564.

Engelsma, G. 1970. *Planta* **91**:246–254.

Engelsma, G. 1972. *Plant Physiol.* **50**:599–602.

Engelsma, G., and J. M. H. van Bruggen. 1971. *Plant Physiol.* **48**:94–96.

Fraenkel, G. S. 1959. *Science* **129**:1466.

Freudenberg, K., and A. C. Neish. 1968. "Constitution and Biosynthesis of Lignin." Springer-Verlag, Berlin and New York.

Gamborg, O. L. 1966. *Can. J. Biochem.* **44**:791–799.

Gibson, F., and J. Pittard. 1968. *Bacteriol. Rev.* **32**:465–492.

Gilchrist, D. G., T. S. Woodin, M. L. Johnson, and T. Kosuge. 1972. *Plant Physiol.* **49**:52–57.

Goldstein, L. D., P. H. Jennings, and H. V. Marsh. 1972. *Plant Cell Physiol.* **13**:783–793

Grisebach, H. 1972. *Hoppe-Seyler's Z. Physiol. Chem.* **353**:123–137.

Hachtel, W., and B. Schwemmle. 1972. *Z. Pflanzenphysiol.* **68**:127–133.

Hahlbrock, K., J. Ebel, R. Ortmann, A. Sutter, E. Wellmann, and H. Grisebach. 1971. *Biochim. Biophys. Acta* **244**:7.

Hanson, K. R., and E. A. Havir. 1972. *Recent Advan. Phytochem.* **4**:45–85.

Hanson, K. R., and M. Zucker. 1963. *J. Biol. Chem.* **238**:1105.

Harborne, J. B., ed. 1964. "Biochemistry of Phenolic Compounds." Academic Press, New York.

Harborne, J. B., ed. 1970. "Phytochemical Phylogeny." Academic Press, New York.

Harborne, J. B., and J. J. Corner. 1961. *Biochem. J.* **81**:242–250.

Havir, E. A., and K. R. Hanson. 1968. *Biochemistry* **7**:1904–1914.

Havir, E. A., and K. R. Hanson. 1973. *Biochemistry* **12**:1583–1591.

Havir, E. A., P. D. Reid, and H. V. Marsh. 1971. *Plant Physiol.* **48**:130–136.

Hess, D., and R. Endress. 1973. *Z. Pflanzenphysiol.* **68**:441–449.

Hess, S. L., and L. A. Hadwiger. 1971. *Plant Physiol.* **48**: 197–202.

Hyodo, H., and S. F. Yang. 1971. *Arch. Biochem. Biophys.* **143**:338–339.

Iredale, S. E., and H. Smith. 1973. *Phytochemistry* **12**:2145–2154.

Katunuma, N., K. Kito, and E. Kominami. 1971a. *Biochem. Biophys. Res. Commun.* **45**:76–81.

Katunuma, N., E. Kominami, and S. Kominami. 1971b. *Biochem. Biophys. Res. Commun.* **45**:70–75.

Klein-Eude, D., P. Rollin, and C. Huault. 1974. *Plant. Sci. Lett.* **2**:1–8.

Koch, G. L. E., D. C. Shaw, and F. Gibson. 1972. *Biochim. Biophys. Acta* **258**:719–730.

Kojima, M., T. Minamikawa, and I. Uritani. 1969. *Plant Cell Physiol.* **10**:245–257.

Koukol, J., and E. E. Conn. 1961. *J. Biol. Chem.* **236**:2692–2698.

Laanest, L. E., and U. V. Margna. 1972. *Fiziol. Rast.* **19**:1157–1164.

Majak, W., and G. H. N. Towers. 1973. *Phytochemistry* **12**:2189–2195.

Mansell, R. L., and M. H. Zenk. 1972. *Z. Pflanzenphysiol.* **68**:286–288.

Minamikawa, T., M. Kojima, and I. Uritani. 1966a. *Arch. Biochem. Biophys.* **117**:194–195.

Minamikawa, T., M. Kojima, and I. Uritani. 1966b. *Plant Cell Physiol.* **7**:583–591.

Minamikawa, T., and I. Uritani. 1965. *J. Biochem. (Tokyo)* **58**:53–59.

Minamikawa, T., I. and Uritani. 1967. *J. Biochem. (Tokyo)* **61**:367–372.
Monod, J., J. P. Changeux, and F. Jacob. 1963. *J. Mol. Biol.* **6**:306–329.
Nambudiri, A. M. D., C. P. Vance, and G. H. N. Towers. 1973. *Biochem. J.* **134**:891–897.
Nari, J., C. Mouttet, M. H. Pinna, and J. Ricard. 1972. *FEBS Lett.* **23**:220–224.
Neish, A. C. 1961. *Phytochemistry* **1**:1–24.
Pressey, R. 1967. *Plant Physiol.* **42**:1780–1786.
Pressey, R., and R. Shaw. 1966. *Plant Physiol.* **41**:1657–1661.
Rebello, J. L., and R. A. Jensen. 1970. *J. Biol. Chem.* **245**:3738–3744.
Riov, J., S. P. Monselise, and R. S. Kahan. 1969. *Plant Physiol.* **44**:631–635.
Russell, D. W. 1971. *J. Biol. Chem.* **246**:3870–3878.
Russell, D. W., and E. E. Conn. 1967. *Arch. Biochem. Biophys.* **122**:256–258.
Sacher, J. A., G. H. N. Towers, and D. D. Davies. 1972. *Phytochemistry* **11**:2383–2391.
Sanderson, G. W. 1966. *Biochem. J.* **98**:248–252.
Scherf, H., and M. H. Zenk. 1967. *Z. Pflanzenphysiol.* **57**:401–418.
Schimke, R. T. 1969. *Curr. Top. Cell. Regul.* **1**:77–124.
Schimke, R. T., E. W. Sweeney, and C. M. Berlin. 1965. *J. Biol. Chem.* **240**:322–331.
Schmitz, M., and U. Seitz. 1972. *Z. Pflanzenphysiol.* **68**:259.
Schopfer, P., and B. Hock. 1971. *Planta* **96**:248–253.
Stafford, H. 1966. *Plant Physiol.* **41**:953–961.
Stafford, H. 1969. *Phytochemistry* **8**:743–752.
Steiner, A. M. 1972. *Z. Pflanzenphysiol.* **68**:266–271.
Swain, T., ed. 1966. "Comparative Phytochemistry." Academic Press, New York.
Swain, T., and C. A. Williams. 1970. *Phytochemistry* **9**:2115–2122.
Taylor, A. D., and M. Zucker. 1966. *Plant Physiol.* **41**:1350–1359.
Walton, D. C., and E. Sondheimer. 1968. *Plant Physiol.* **43**:467–469.
Weber, H. L., and A. Böck. 1969. *Arch. Mikrobiol.* **66**:250–258.
Weidner, M., I. Rissland, L. Lohmann, C. Huault, and H. Mohr. 1969. *Planta* **86**:33–41.
Wiermann, R. 1973. *Planta* **110**:353–360.
Woodin, T., and T. Kosuge. 1968. *Plant Physiol.* **43**:5–47.
Zucker, M. 1963. *Plant Physiol.* **38**:575–580.
Zucker, M. 1965. *Plant Physiol.* **40**:779–784.
Zucker, M. 1969. *Plant Physiol.* **44**:912–922.
Zucker, M. 1970. *Biochim. Biophys. Acta* **208**:331–333.
Zucker, M. 1971. *Plant Physiol.* **47**:442–444.
Zucker, M. 1972. *Annu. Rev. Plant Physiol.* **23**:133–156.
Zucker, M., and C. C. Levy. 1959. *Plant Physiol.* **34**:108–112.

ENZYMOLOGY AND REGULATION OF FLAVONOID AND LIGNIN BIOSYNTHESIS IN PLANTS AND PLANT CELL SUSPENSION CULTURES

H. GRISEBACH and K. HAHLBROCK

Biological Institute II, University of Freiburg i. Br., Germany

Introduction

Extensive tracer studies on flavonoid and lignin biosynthesis with intact plants or plant tissues have led to a basic knowledge of the building units and to an understanding of some details of their biosynthesis. This work has been covered in several recent reviews (Neish, 1964; Grisebach, 1965, 1967, 1968; Grisebach and Barz, 1969; Pachéco, 1969; Freudenberg and Neish, 1968; Sarkanen and Ludwig, 1971). As in studies of other biosynthetic pathways, however, it became apparent that a more detailed knowledge of the nature and sequence of the individual biosynthetic steps and their regulation could only be gained by investigations of the enzymes involved.

Work on the enzymology of phenylpropane metabolism commenced with the discovery in 1961 of the enzyme phenylalanine ammonia-lyase (PAL) by Koukol and Conn. Further progress in this field was rather slow for some years, probably because of relatively low concentrations and variations in the activities of enzymes of secondary metabolism during most stages of plant growth (see next section). A major breakthrough came with the use of plant cell suspension cultures as a source of enzymes, which led in short succession to the discovery of a number of enzymes of flavonoid biosynthesis. Very recently some progress has also been made in the enzymology of the biosynthesis of lignin precursors.

Since we have recently written an extensive review on the biosynthesis of flavonoids including the enzymology (Hahlbrock and Grisebach, 1973), we will describe mainly our own work in this chapter.

Enzymology of Flavonoid Glycoside Formation in *Petroselinum hortense* and Comparison with Enzymes from Other Plants

INTRODUCTION

Most investigations on the enzymology of flavonoid biosynthesis have been carried out with parsley seedlings or cell suspension cultures of this plant. Emphasis will therefore be placed on the work with parsley, but a comparison with the corresponding enzymes from other plants will be presented. Our work with parsley started in 1964 when we became interested in the biosynthesis of the branched-chain sugar apiose occurring in this plant as a glycosidic component of apiin (see Fig. 15) (Grisebach and Döbereiner, 1964). At that time we did not know that parsley would become our "Hauspflanze" for the next 9 years.

Our first attempts to find enzymes of phenylpropanoid metabolism in parsley were not very successful. In our search for a cinnamic acid:CoA ligase, we found only very low activation of some cinnamic acids in the presence of coenzyme A, ATP, and a protein preparation from parsley leaves, and it was not clear whether this activation had anything to do with phenylpropanoid metabolism (Grisebach *et al.*, 1966). Only later did it become apparent that the plants we had used at first were much too old. When the relationship between organ development and activity of some enzymes involved in flavone glycoside biosynthesis was studied in young parsley plants, a sharp and progressive decline in specific activities was observed in extracts made from progressively older organs (Fig. 1a) (Hahlbrock *et al.*, 1971a). Comparison of the specific activities of several of these enzymes with the concentration of flavone glycosides in the respective organs at various stages of development revealed a substantial correlation (Fig. 1b). However, more detailed studies on this regulatory aspect and, equally important, the isolation of amounts of protein sufficient to permit purification of enzymes, were limited by the minute size of the young leaves. It was therefore good luck that one of us (K. H.) was able to bring back from a visit to Madison, Wisconsin, a piece of parsley callus tissue from Prof. A. C. Hildebrandt, which was used to start a liquid suspension culture.

It was soon found that these cultures produce flavone glycosides upon illumination with white light. Two major glycosides formed were identified

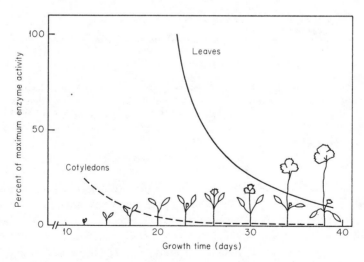

Fig. 1a. Relative changes in the activities of phenylalanine ammonia-lyase, chalcone-flavanone isomerase, UDP-glucose:flavonoid 7-*O*-glucosyltransferase, and UDP-apiose:flavone glycoside apiosyltransferase in cotyledons (- - -) and leaves (——) of developing parsley plants.

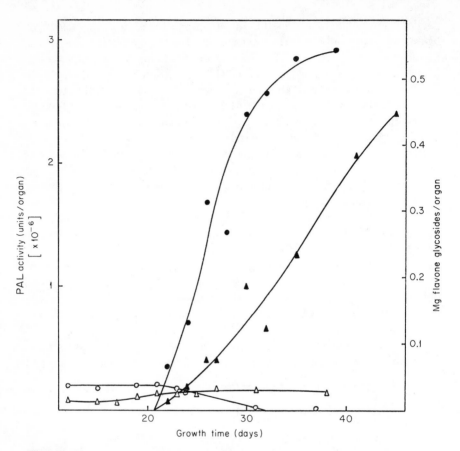

Fig. 1b. Changes in phenylalanine ammonia-lyase (PAL) activity and accumulation of flavone glycosides per organ in developing parsley plants. All of the other enzymes mentioned in the legend to Fig. 1a gave curves of identical shape and can be substituted for PAL. Cotyledons: ○—○, PAL; △—△, flavone glycosides. Leaves: ●—●, PAL; ▲—▲, flavone glycosides.

as apiin and graveobiosid B (3′-methoxyapiin) (see Fig. 6). Changes in the activities of PAL and UDP-apiose synthase could also be measured easily with these cultures over a period of several days, and it was found that the two enzymes reached maximum specific activities after different periods of time followed by a rapid decline (Fig. 2) (Hahlbrock and Wellmann, 1970). The long-sought cinnamic acid:CoA ligase could also be detected in these cultures, and in this case also the extractable enzyme ac-

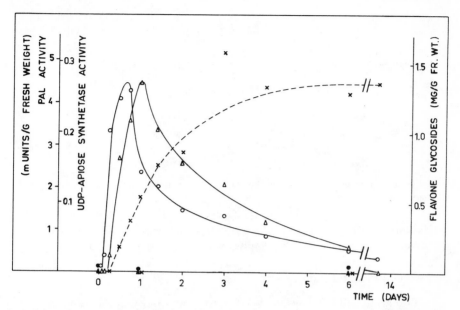

Fig. 2. Changes in phenylalanine ammonia-lyase (PAL) and UDP-apiose synthase activity and accumulation of flavone glycosides in cell suspension cultures of parsley during illumination with white light. PAL: O—O, light; ●—●, dark. UDP-apiose synthase: △—△, light; ▲—▲, dark. Flavone glycosides: ✕- - -✕.

tivity was markedly increased by illumination of the cell cultures (Hahlbrock and Grisebach, 1970). These experiments marked the beginning of our studies with cell cultures which have led to the detection of all 13 enzymes postulated for the biosynthesis of apiin and graveobioside B in parsley.

STRUCTURE OF FLAVONOID GLYCOSIDES FROM ILLUMINATED CELL SUSPENSION CULTURES OF PARSLEY

For the enzymatic work it was important to know the nature of the flavonoids which are produced by the parsley cultures. A total of 24 different flavonoid glycosides were isolated from the cell cultures after continuous illumination for 24 hours. The chemical structures of 14 of these compounds were established. The products formed were malonylated and nonacylated flavone 7-*O*-glucosides and 7-*O*-apiosyl(1→2)-glucosides, and flavonol 7-*O*-glucosides and 3,7-*O*-diglucosides (Table 1) (Kreuzaler and

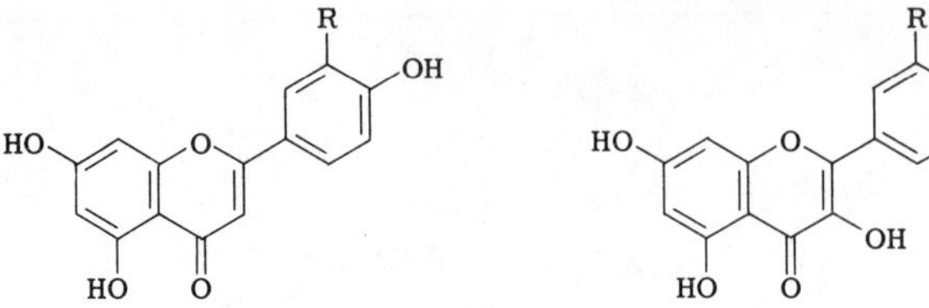

Flavones:	R =	Flavonols:
Apigenin	H	Kaempferol
Luteolin	OH	Quercetin
Chrysoeriol	OCH_3	Isorhamnetin

FIG. 3. Aglycones of flavone and flavanol glycosides formed upon illumination of cell suspension cultures of *Petroselinum hortense*.

Hahlbrock, 1973). Figure 3 illustrates the close structural similarities between the flavone and flavonol aglycones.

TWO GROUPS OF ENZYMES IN PARSLEY

Two groups of enzymes involved in flavonoid biosynthesis could be distinguished on the basis of concomitant changes in their activities following

TABLE 1

CHARACTERIZATION OF SOME OF THE FLAVONE AND FLAVONOL GLYCOSIDES ISOLATED
FROM ILLUMINATED CELL SUSPENSION CULTURES OF PARSLEY

Aglycone	Glucose	Apiose	Position of glycosylation	Electrophoretic mobility	Malonic acid	Molar ratio, aglycone: glucose:apiose
Apigenin	+	−	7	−	−	1:1·2
Apigenin	+	−	7	+	+	1:1·2
Apigenin	+	+	7	−	−	1:1·3:0·8
Apigenin	+	+	7	+	+	1:1·2:0·9
Luteolin	+	−	7	−	−	1:1·3
Chrysoeriol	+	−	7	−	−	1:0·8
Chrysoeriol	+	−	7	+	+	1:0·9
Chrysoeriol	+	+	7	−	−	1:1·0:1·0
Chrysoeriol	+	+	7	+	+ (?)	1:1·1:1·1
Quercetin	+	−	3	−	−	1:1·3
Quercetin	+	−.	3 + 7	−	−	1:2·2
Quercetin	+	−	3	+	+	n.d.[a]
Isorhamnetin	+	−	3 + 7	+	+	1:2·1
Isorhamnetin	+	−	3 + 7	++	+	1:1·7

[a] n.d. = not determined.

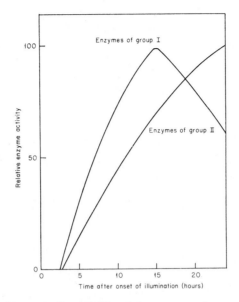

FIG. 4. Relative changes in the activities of the enzymes of general phenylpropanoid metabolism (group I) and of flavone glycoside biosynthesis (group II), following the illumination of cell suspension cultures of *Petroselinum hortense*. For details see sections on Two Groups of Enzymes in Parsley, and Further Aspects of Regulation of Flavonoid Glycoside Biosynthesis.

the illumination of previously dark-grown cells (Fig. 4). The first group consists of the enzymes of general phenylpropanoid metabolism, while the second group comprises all of the enzymes specifically related to flavonoid biosynthesis (Hahlbrock *et al.*, 1971b; Hahlbrock, 1972). As a consequence of this observation, most of the previously unknown enzymes of the flavonoid glycoside pathway could be isolated from illuminated parsley cell suspension cultures. Apart from the substrate specificities of these enzymes, the characteristic light-induced changes in activity were regarded as strong evidence for the *specific* function of these enzymes in flavonoid biosynthesis.

The enzymes which catalyze the sequence of reactions involved in the conversion of phenylalanine and/or tyrosine to activated cinnamic acids are phenylalanine ammonia-lyase (PAL), cinnamic acid 4-hydroxylase, *p*-coumarate:CoA ligase, phenolase(s), and methyltransferase(s) (Fig. 5). The products of this biosynthetic sequence are not only intermediates in flavonoid biosynthesis, but also precursors of a number of other phenolic compounds in higher plants, notably lignin. The sequence of enzymatic steps involved in the formation of malonylated apigenin 7-O-(β-D-apio-furanosyl(1→2)β-D-glucoside (malonylapiin), one of the major flavonoid

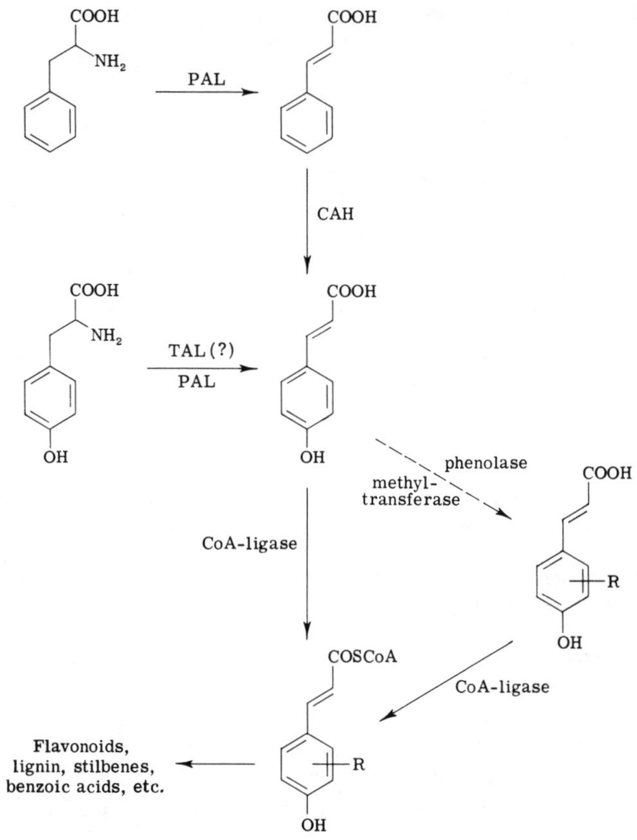

FIG. 5. Reactions catalyzed by the enzymes of general phenylpropanoid metabolism. TAL, tyrosine ammonia-lyase; CAH, cinnamic acid 4-hydroxylase; R, hydroxy and/or methoxy groups in various positions.

constituents in parsley, from *p*-coumaroyl-CoA and malonyl-CoA is shown in Fig. 6. Some properties of these enzymes, according to our present knowledge, are summarized in Table 2.

Phenylalanine Ammonia-Lyase

The light-induced enzyme from parsley cell cultures was purified to apparent homogeneity as judged from disk gel electrophoresis (A. Zimmermann and K. Hahlbrock, unpublished results, 1973). Tyrosine ammonia-lyase activity was also associated with the enzyme throughout the entire purification procedure. A molecular weight of about 3×10^5 daltons was estimated by gel chromatography, indicating that the molecular weight of the enzyme from parsley is very similar to that of phenylalanine ammonialyases from maize and potato (Havir and Hanson, 1973).

Two different methods from the stimulation of phenylalanine ammonialyase activity in parsley cells have been reported (Hahlbrock *et al.*, 1971b; Hahlbrock and Wellmann, 1973). Possible implications of these observations as to the molecular mechanism which causes the observed changes

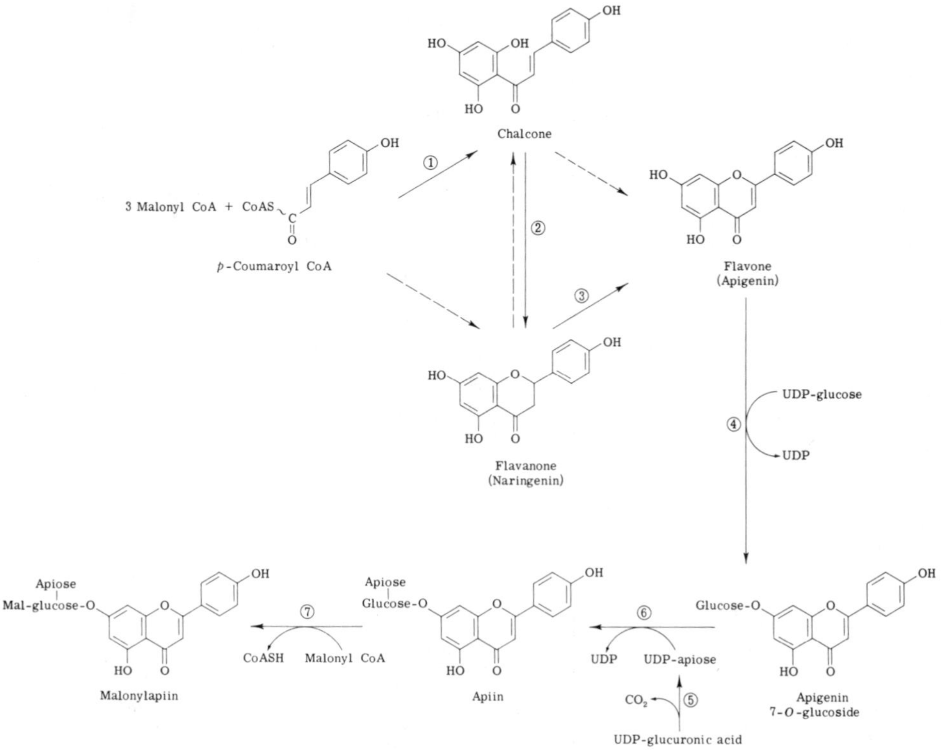

FIG. 6. Proposed scheme for the enzymatic formation of apigenin glycosides in illuminated cell suspension cultures of *Petroselinum hortense*. The enzymes involved in this sequence of reactions are (1) chalcone-flavanone synthetase, (2) chalcone-flavanone isomerase, (3) chalcone-flavanone oxidase, (4) UDP-glucose:flavone/flavonol 7-*O*-glucosyltransferase, (5) UDP-apiose synthase, (6) UDP-apiose:flavone 7-*O*-glucoside apiosyltransferase, (7) malonyl-CoA:flavonoid glycoside malonyltransferase. A possible alternative route for reactions 1 to 3 is indicated by dotted arrows.

TABLE 2

ENZYMES OF FLAVONOID GLYCOSIDE BIOSYNTHESIS FROM ILLUMINATED CELL SUSPENSION CULTURES OF *Petroselinum hortense*

Enzyme	Degree of purification	Molecular weight (daltons)	pH optimum	Substrate specificity	Apparent K_m values of substrates	Remarks	References
Chalcone synthetase	Crude extract	n.d.[a]	8[b]	—	—	—	Kreuzaler and Hahlbrock (1972)
Chalcone-flavanone isomerase	Crude extract	n.d.	7[b]	Specific for phloroglucinol-type substitution pattern of ring A[b]	—	—	Hahlbrock et al. (1970a)
UDP-Glucose: flavone/flavonol 7-O-glucosyltransferase	89-Fold	50,000	7.5	7-O-Glucosylation of various flavones, flavonols, flavanones; also conversion of quercetin 3-O-glucoside[b]	Apigenin ($2.7 \times 10^{-6}\ M$), luteolin ($1.5 \times 10^{-6}\ M$), naringenin ($1 \times 10^{-5}\ M$), UDP-glucose ($1.2 \times 10^{-4}\ M$), TDP-glucose ($2.6 \times 10^{-4}\ M$)	—	Sutter et al. (1972)
UDP-Glucose: flavonol 3-O-glucosyltransferase	40-Fold[a]	50,000	8	3-O-Glucosylation of various flavonols incl. quercetin 7-O-glucoside; no conversion of dihydroquercetin	Quercetin ($<10^{-6}\ M$)	Complete separation of the two glucosyltransferases on DEAE-cellulose	Sutter and Grisebach (1973)

Enzyme	Purification	Molecular weight	pH optimum	Properties	K_m values	Remarks	Reference
UDP-Apiose synthase	1000-Fold	101,000–115,000	8.2	—	UDP-D-glucuronic acid (2×10^{-6} M)	Exhibits also UDP-xylose synthetase activity	Baron et al. (1972, 1973)
UDP-Apiose: flavonoid 7-O-glucoside apiosyltransferase	123-Fold	55,000	7.0	Apiosylation of 7-O-glucosides of various flavones, flavanones, isoflavones; no conversion of flavonol 7-O-glucosides	Apigenin 7-O-glucoside (6.6×10^{-6} M)	—	Ortmann et al. (1972)
Malonyl-CoA: flavonoid glycoside malonyltransferase	Crude extract	n.d.	n.d.	—	—	—	Hahlbrock (1972)
SAM: o-dihydric phenol m-O-methyltransferase	82-Fold	48,000	9.7	Methylation in the *meta* position of various flavonoids, caffeic acid, protocatechuic acid	Luteolin (4.6×10^{-5} M), luteolin 7-O-glucoside, (3.1×10^{-5} M), eriodictyol (1.2×10^{-3} M), caffeic acid (1.6×10^{-3} M)	—	Ebel et al. (1972)

[a] n.d. = Not determined.
[b] Unpublished results.

in enzyme activity are discussed at below in the section on Further Aspects of Regulation of Flavonoid Glycoside Biosynthesis. Work on PAL from other plants has recently been reviewed by Hanson and Havir (1972a,b) and by Camm and Towers (1973).

Cinnamic Acid 4-Hydroxylase

As in the case of the cinnamic acid 4-hydroxylase from *Pisum sativum* (Russell and Conn, 1967; Russell, 1971), the enzyme from parsley cell cultures is a microsomal mixed-function oxidase which requires molecular oxygen, NADPH, and mercaptoethanol for activity (Hahlbrock *et al.*, 1971b; Büche, 1973; Büche and Sandermann, 1973). The pH optimum is around pH 7.5 and the apparent K_m value for cinnamic acid in 0.1 M phosphate buffer is 1.7 $\times$ 10^{-5} M. After removal of about 60 percent of the microsomal phospholipid, cinnamic acid 4-hydroxylase activity was completely abolished. Reactivation of enzyme activity could be achieved by addition of a crude microsomal lipid extract (Büche and Sandermann, 1973).

p-Coumarate:CoA Ligase

p-Coumarate:CoA ligase was first isolated from illuminated cell suspension cultures of parsley (Hahlbrock and Grisebach, 1970). The formation of coenzyme A thiol esters of cinnamic, p-coumaric, p-methoxycinnamic, and ferulic acids was catalyzed by the crude enzyme preparation. Whereas the formation of acetyl-CoA was not significantly influenced by light, enzyme activity for the activation of p-coumaric acid was markedly increased (see Fig. 4). Subsequently, the enzyme was also demonstrated in cell suspension cultures of soybean (*Glycine max*) during a short period of the growth cycle which was characterized by concomitant large increases in the activities of PAL and p-coumarate:CoA ligase (Hahlbrock *et al.*, 1971c; Hahlbrock and Kuhlen, 1972).

An enzyme preparation from soybean cell suspension cultures was first purified about 12-fold and used as a convenient source for the enzymatic synthesis of CoA thiol esters (Lindl *et al.*, 1973). The partially purified enzyme was stable for several months when kept frozen at −20°C. At temperatures above 0°C, however, the activity of the enzyme decreased rapidly in buffered aqueous solution. This loss of enzyme activity could be prevented by the addition of 30 percent glycerol to the buffer. With this precaution, extensive further purification was achieved, resulting in a separation of the ligase activity into two enzymes which differed with respect to their elution patterns from ion-exchange columns and in their substrate specificities. The two enzymes were purified approximately

150-fold ("ligase I") and 50-fold ("ligase II"), respectively (Knobloch and Hahlbrock, 1973). The two ligases were recovered from a gel column with identical amounts of elution buffer, suggesting that their molecular weights are very similar. Using a mixture of the two enzymes the molecular weights were estimated to be about 55,000 daltons (Lindl *et al.*, 1973).

PHENOLASE

Although a large amount of work on plant phenolases has been done, there is very little information on the significance of this type of enzyme for flavonoid biosynthesis. A phenolase preparation catalyzing the hydroxylation of *p*-coumaric acid to caffeic acid with ascorbate as reducing agent was obtained from parsley cell cultures (Schill and Grisebach, 1973). In contrast to the phenolase from *Beta vulgaris* (see below), the enzyme preparation from parsley cells showed only weak hydroxylating activity with naringenin as substrate and no detectable reaction with dihydrokaempferol or apigenin. Whereas light increased the activity of all enzymes involved in flavone glycoside biosynthesis in parsley (see section on two groups of enzymes in parsley, above), it had no effect on the extractable enzyme activity for hydroxylation of *p*-coumaric acid or naringenin. Therefore, it seems unlikely that this extractable phenolase is specifically involved in flavone glycoside biosynthesis in parsley cell cultures.

A purified spinach beet (*Beta vulgaris*) phenolase catalyzed the hydroxylation of *p*-coumaric acid to caffeic acid (Vaughan and Butt, 1969, 1970) as well as hydroxylation of 4'-hydroxy flavonoids in the 3'-position (Vaughan *et al.*, 1969; Roberts and Vaughan, 1971). With *p*-coumaric acid, kaempferol, dihydrokaempferol, and naringenin as substrates, K_m values were 6.4×10^{-4}, 22×10^{-4}, 52×10^{-4}, and 1.6×10^{-4} M, respectively, and V_{max} values 1.67, 9, 242, and 8 nmole/min per m-unit enzyme, respectively (Roberts and Vaughan, 1971; P. F. T. Vaughan, personal communication). However, since the nature and concentration of flavonoids present in spinach beet is unknown, it is not possible to judge the significance of this phenolase for flavonoid biosynthesis. It seems possible that hydroxylases of the type found in microorganisms (Neujahr and Gaal, 1973) might take part in flavonoid biosynthesis, and one should examine higher plants for such enzymes.

HYDROXYCINNAMIC ACID 3-*O*-METHYLTRANSFERASE

Specific methylation in the 3-*O*-position of various hydroxycinnamic acids in cell-free extracts from plants has been reported by Finkle and Nelson (1963), Finkle and Masry (1964), Hess (1966), Shimada *et al.*,

(1970, 1972), Mansell and Seder (1971), and Ebel *et al.*, (1972). The donor of the methyl groups was shown to be *S*-adenosylmethionine (Finkle and Masry, 1964). While most of these reactions were not specific for cinnamic acids but also took place with a number of other substrates, including several flavonoid compounds, Hess (1966) based his "cinnamic acid starter hypothesis" partially on the observation that cinnamic acids were more efficient substrates for 3-*O*-methylation in extracts from *Petunia hybrida* than anthocyanins. By contrast, however, recent studies on the substrate specificity of an *O*-methyltransferase purified from cell suspension cultures of parsley (Ebel *et al.*, 1972) and on changes in the activity of this enzyme after treatment of the cells with light, suggested that in this case methylation takes place preferentially at the flavonoid stage (see methyltransferase section, below).

A final answer to the question at which stages do hydroxylation and methylation reactions occur during flavonoid biosynthesis will be dependent upon the results of studies on the substrate specificities of all of the enzymes related to this pathway, especially those of *p*-coumarate:CoA ligase and chalcone-flavanone synthetase, the two enzymes involved in reactions at the interlock of general phenylpropanoid metabolism and flavonoid biosynthesis.

Chalcone-Flavanone Synthetase

As an extension of Birch's hypothesis (Birch and Donovan, 1953), it was postulated (Grisebach, 1962) that the first reaction specific for the biosynthesis of flavonoid compounds is the enzyme-mediated condensation of an activated cinnamic acid with three molecules of malonyl-CoA to give a chalcone (Fig. 7). This suggestion was substantiated by feeding experiments with (3-^{14}C)-cinnamic acid using shoots of *Cicer arietinum* from which labeled 4,4′,6′-trihydrochalcone could be isolated by dilution analysis (Grisebach and Brandner, 1962). More recently, Endress (1972) isolated radioactive 3,4,2′,4′,6′-pentahydroxychalcone after administration of (1-^{14}C)-acetate to buds of *Petunia hybrida*.

Direct evidence for the reaction mechanism formulated in Fig. 8 has been obtained from experiments with an enzyme preparation from cell suspension cultures of parsley which catalyzes the formation of naringenin (5,7,4′-trihydroxyflavanone) from *p*-coumaroyl-CoA and malonyl-CoA (Kreuzaler and Hahlbrock, 1972). While no experimental evidence has been obtained so far regarding possible intermediates in the formation of the aromatic ring A from "acetate units," chemical degradation of the overall reaction product proved that ^{14}C-labeled malonyl-CoA was incorporated exclusively into this ring (Fig. 8) (Kreuzaler and Hahlbrock,

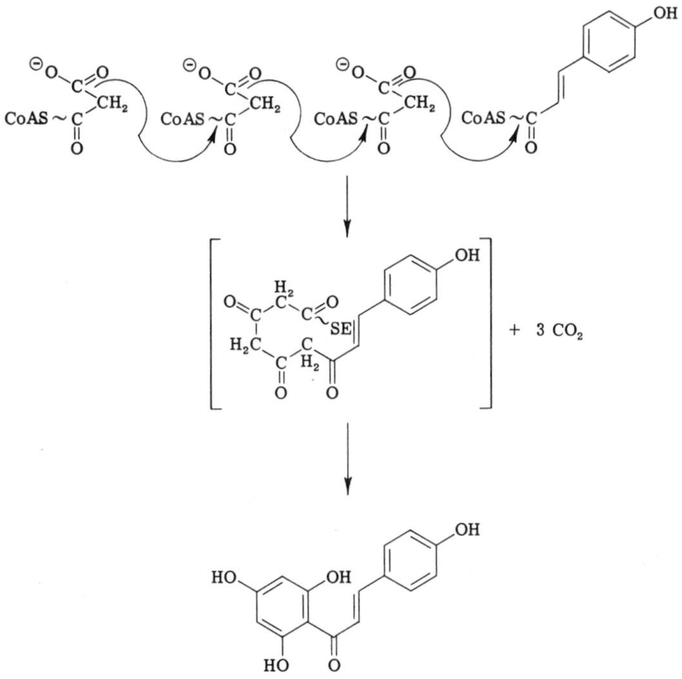

FIG. 7. Hypothetical scheme for the formation of 4,2′,4′,6′-tetrahydroxychalcone from *p*-coumaroyl-CoA and malonyl-CoA.

1972). Phenolic compounds which might possibly interfere with chalcone-flavanone synthetase activity were removed from buffered cell extracts with an anion exchanger. The enzyme has an optimum around pH 8 and does not require the addition of any cofactors for activity (F. Kreuzaler, unpublished). Since chalcone-flavanone synthetase has not been obtained free of chalcone-flavanone isomerase activity, it cannot be decided at present whether a chalcone or flavanone is the immediate product of this enzyme.

In analogy to the biosynthesis of 6-methylsalicyclic acid (Dimroth *et al.*, 1970), it can be postulated that the reactions leading to chalcone/flavanone occur on a multienzyme complex. At least four enzyme functions must be postulated for such a complex (Fig. 9): transacetylases for *p*-coumaryl-CoA (1) and malonyl-CoA (2), a condensing enzyme (3), and an enzyme (4) catalyzing the formation of ring A and the release of the final product, either the chalcone or the flavanone.

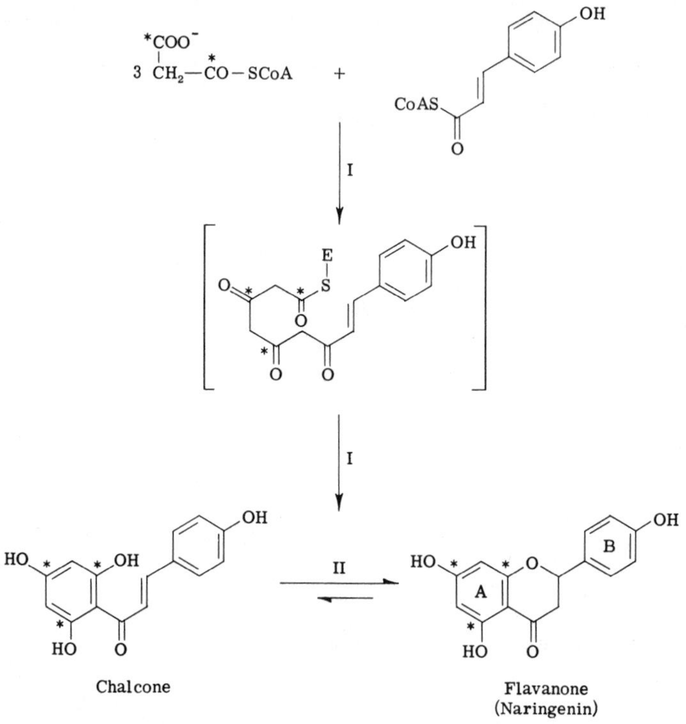

FIG. 8. Proof for the incorporation of [1-¹⁴C]-malonyl-CoA into ring A of naringenin. I, Reactions catalyzed by chalcone synthetase; II, reaction catalyzed by chalcone-flavanone isomerase; E, enzyme. Naringenin was further degraded via apigenin to phloroglucinol, *p*-hydroxybenzoic acid, and *p*-hydroxyacetophenone. All radioactivity was located in phloroglucinol.

CHALCONE-FLAVANONE ISOMERASE

Chalcone-flavonone isomerase, first discovered in *Cicer arietinum* L. by Wong and Moustafa (1966), catalyzes the formation of the six-membered heterocyclic ring of flavanones from the corresponding chalcones (see Fig. 6). There is no unequivocal evidence as yet concerning the direction of the enzyme-mediated isomerization *in vivo* (Grisebach and Barz, 1969).

The enzyme prepared from parsley leaves has been studied more closely. The parsley isomerase has a high substrate specificity for A-ring substitution (flavanone numbering). Of the seven chalcones tested, only 2′,4,4′,6′-

tetrahydroxychalcone with a phloroglucinol type of substutition in ring A was found to serve as substrate (Hahlbrock *et al.*, 1970a). This is consistent with the fact that only flavonoids of this substitution pattern in ring A have been found in parsley.

The K_m values measured for several chalcones with isomerases from other plants suggest that substrate specificities of chalcone-flavanone isomerases in general correspond significantly to the substitution patterns of flavonoid compounds occurring in any particular plant (Table 3 and Fig. 10). The finding that the isomerases from *Phaseolus aureus* and *Cicer arietinum*

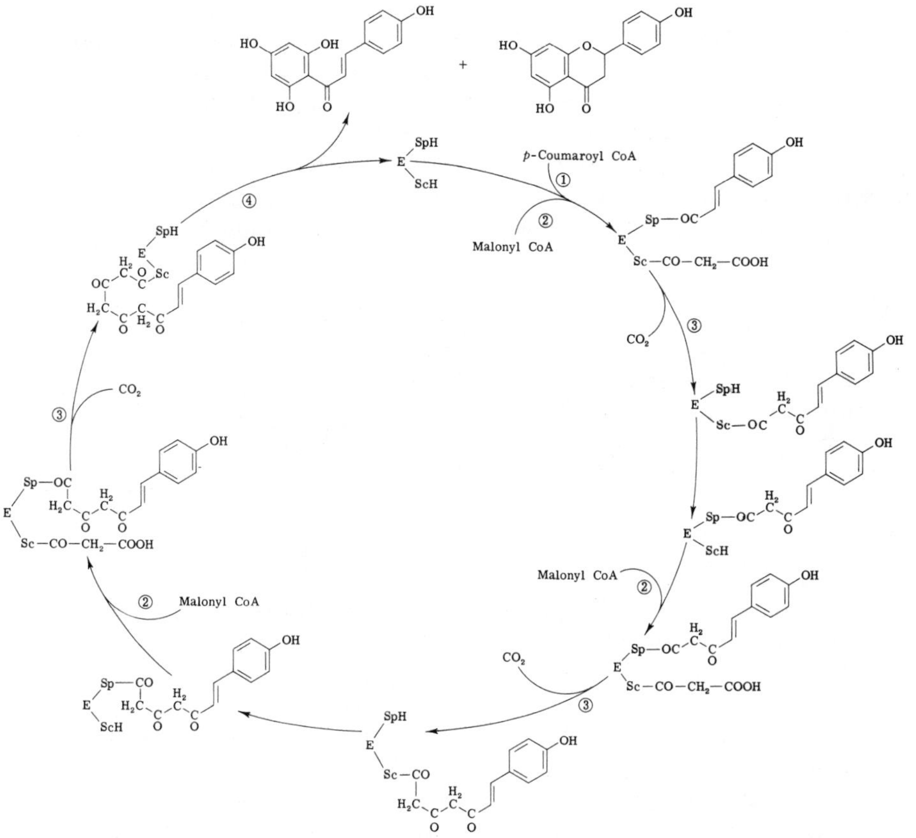

FIG. 9. Hypothetical scheme for a multienzyme complex of chalcone/flavanone synthetase. SpH, Peripheric SH group; ScH, central SH group; 1, *p*-coumaroyl transfer; 2, malonyl transfer; 3, condensation reaction; 4, ring closure and release from the enzyme.

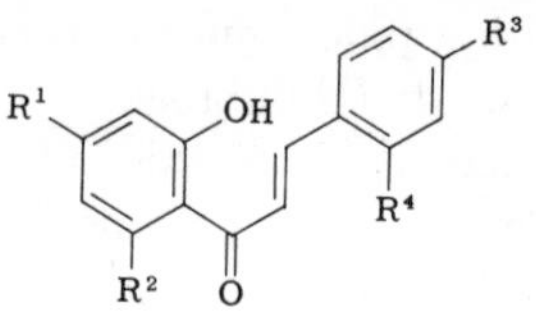

Fig. 10. Chalcones tested as substrates of the chalcone-flavanone isomerases listed in Table 3: (A) R^1 = OH; R^2 = R^3 = R^4 = H. (B) R^1 = R^2 = R^4 = H; R^3 = OH. (C) R^1 = R^3 = OH; R^2 = R^4 = H. (D) R^1 = R^2 = OH; R^3 = R^4 = H. (E) R^1 = R^2 = R^3 = OH; R^4 = H. (F) R^1 = R^2 = R^4 = OH; R^3 = H. (G) R^1 = OGlc; R^3 = OH; R^2 = R^4 = H. (H) R^1 = OGlc; R^2 = OH; R^3 = R^4 = H. (I) R^1 = OGlc; R^2 = R^3 = OH; R^4 = H. (K) R^1 = OGlc; R^2 = R^4 = OH; R^3 = H. (L) R^2 = OGlc; R^1 = R^3 = OH; R^4 = H.

have rather low specificities is consistent with the fact that both phloro-glucinol- and resorcinol-type substituted flavonoid compounds occur in these plants. These conclusions are further supported by tracer studies which suggest that the removal of the 5-hydroxy group (flavanone number-ing) takes place at a stage prior to chalcone formation (Grisebach and Brandner, 1961). Since none of the chalcone glucosides listed in Table 3 were isomerized by the enzymes tested, it must be assumed that chalcones, and not chalcone glucosides, are the natural intermediates in flavonoid biosynthesis.

Chalcone-flavanone isomerases from all plants so far investigated could be separated into varying numbers of isoenzymes (Hahlbrock *et al.*, 1970a; Grambow and Grisebach, 1971). Differences in the pH optima and Michaelis constants for a number of substrates of the two isomerases purified from mung bean seedlings suggested that these enzymes must be regarded as true isoenzymes and not as enzymes differing only in protein conformation.

The stereochemistry of the isomerase reaction has been investigated. Cyclization of 4,2′,4′-trihydroxychalcone catalyzed by either one of the two isomerases from mung bean seedlings leads to the (−) (2S)-7,4′-dihy-droxyflavanone (Hahlbrock *et al.*, 1970b). The absolute configuration at C-2 was determined by circular dicroism measurements. The results on the stereochemical course of the reaction at C-3 of the flavanone with [α-^{2}H]trihydroxychalcone in H_2O and with trihydroxychalcone in ^2H_2O showed that deuterium was incorporated preferentially into either the equatorial or the axial position at C-3, respectively (Fig. 11). A mechanism of the isomerase reaction consistent with these results would be an acid–base-catalyzed reaction leading to the flav-3-en-4-ol intermediate, which undergoes a stereospecific proton transfer to form the flavanone (Fig. 12) (Hahlbrock *et al.*, 1970b).

TABLE 3

COMPARISON OF SOME CHALCONE-FLAVANONE ISOMERASE ISOENZYMES AND HETEROENZYMES

Source of enzyme	Number of isoenzymes		Chalcones tested (see Fig. 10)	Chalcones converted (K_m values, mole/liter)	Basis for distinction between isoenzymes	Reference
	Separated	Tested				
Phaseolus aureus (seedlings)	2	2	Isoenzyme I: A, B, C, E, G, I, L Isoenzyme II: A, B, C, E, G, I, L	A(2.6×10^{-4}), B(1.9×10^{-4}), C(5.7×10^{-5}), E(1.8×10^{-5}) A(4.1×10^{-5}), B(3.7×10^{-5}), C(1.4×10^{-5}), E(4.4×10^{-5})	Differences in electrophoretic mobility, K_m values, pH optimum	Hahlbrock *et al.* (1970a)
Cicer arietinum (germs)	$\geqslant 2$	1	A, B, C, E, I, L	A(1.8×10^{-5}), B(7×10^{-6}), C(1.6×10^{-5}), E(4.4×10^{-5})	Differences in electrophoretic mobility	Hahlbrock *et al.* (1970a)
Petroselinum hortense (leaves)	$\geqslant 5$	1	A, B, C, E, G, I, L, L	E(1.6×10^{-5})	Differences in electrophoretic mobility	Hahlbrock *et al.* (1970a)
Datisca cannabina (leaves)	$\geqslant 3$	Mixture	C, D, E, F, I, K, L	D, E, F (K_m not det.)	Differences in electrophoretic mobility	Grambow and Grisebach (1971)

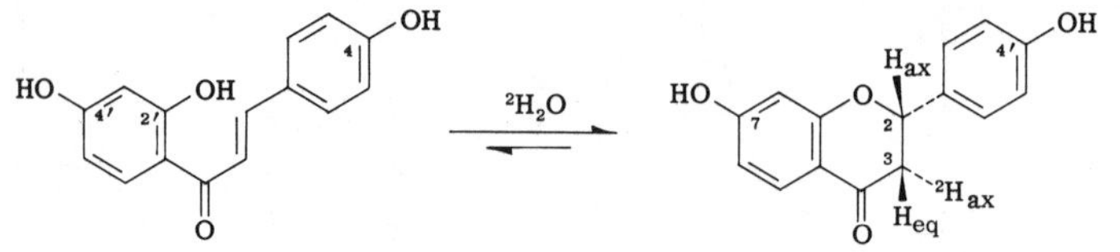

FIG. 11. Stereospecific incorporation of deuterium into 7,4'-dihydroxyflavanone during the chalcone-flavanone isomerase reaction in ^2H_2O.

CHALCONE-FLAVANONE OXIDASE

Chalcone-flavanone oxidase (Fig. 6, enzyme 3) is the only enzyme of apiin biosynthesis which has not yet been detected in cell cultures of parsley. It has been found in young parsley leaves. In a growth period in which other enzymes of apiin biosynthesis show maximum activity (Fig. 1), very young primary leaves of parsley seedlings also contain enzymatic activity for the transformation of naringenin to apigenin (Sutter and Grisebach, 1973). In the extract with Tris-HCl buffer of pH 7.5 containing mercaptoethanol, the enzymatic activity was present in the supernatant after centrifugation at 150,000g for 90 minutes. The reaction has a pH optimum of 7–7.5. More than 50 percent conversion of naringenin to apigenin could be obtained.

The reaction is cofactor-dependent. After gel filtration of the crude extract on a Sephadex G-25 column or after dialysis, enzymatic activity was lost. Reactivation could be achieved by the addition of small amounts of crude extract or heat-denatured extract. One of the cofactors seems to be iron. Addition of 10^{-3} M Fe^{2+} or Fe^{3+} stimulated the reaction in the crude extracts two- to fivefold. Phenanthroline at a concentration of 10^{-3} M inhibited the reaction completely. This inhibition could be compensated by addition of 5×10^{-3} M Fe^{2+}. Besides iron, other cofactors must be required for the reaction, since a G-25 filtrate could not be reactivated by iron alone. Reactivation by addition of flavin and pyridine nucleotides in combination

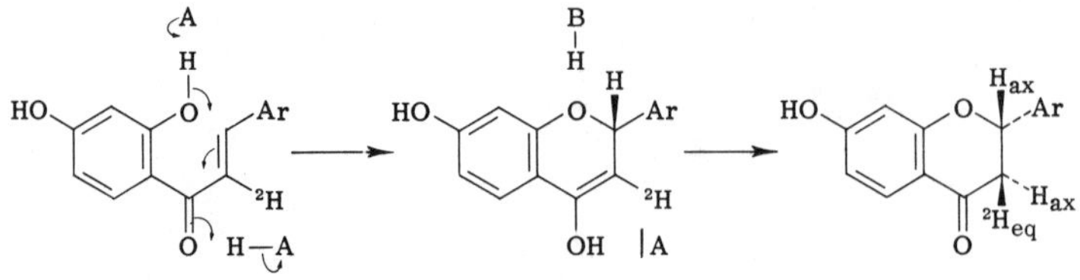

FIG. 12. Hypothetical reaction mechanism for chalcone-flavanone isomerase. B, Basic group on enzyme; A, acidic group on enzyme.

with various one- and two-electron acceptors and metal ions was not successful.

According to our data, the enzyme catalyzing the oxidation of naringenin or naringenin-chalcone to apigenin cannot be a peroxidase. Addition of catalase to the incubation mixture had no influence on the yield of apigenin. Addition of 10^{-5} M H_2O_2 had no influence and 10^{-3} M H_2O_2 inhibited the reaction to about 50 percent. When [2-^{14}C]-naringenin was incubated with horseradish peroxidase in the presence of H_2O_2, radioactivity was released as $^{14}CO_2$. An as yet unidentified intermediate of this oxidation could be detected on chromatograms with 30 percent acetic acid (R_f 0.85). This product was not formed when crude parsley extracts were incubated with naringenin in the presence of H_2O_2. Attempts to demonstrate the chalcone-flavanone oxidase activity in extracts from parsley cell cultures during a period when production of apigenin glycosides is occurring have so far failed.

GLUCOSYLTRANSFERASES

Flavone 7-*O*-glycosides as well as flavonol 7-*O*-glycosides and 3,7-di-*O*-glycosides were shown to be formed by illuminated parsley cell suspension cultures (see above, Structure of Flavonoid Glycosides . . .). Correspondingly, two distinct enzymes glucosylating flavonoids specifically in the 7-*O*- and in the 3-*O*-position, respectively, were isolated from these cells. The reactions catalyzed by the two glucosyltransferases are shown in Fig. 13. Both enzymes have a molecular weight of about 55,000 daltons, as calculated from the elution volumes after gel chromatography (Sutter *et al.*, 1972; Sutter and Grisebach, 1973). They could be separated according to their different charge by DEAE-cellulose column chromatography or by disk gel electrophoresis. Neither of the two enzymes requires cofactors, thiol reagents, or divalent cations for activity.

The first enzyme, UDP-glucose:flavone/flavonol 7-*O*-glucosyltransferase, was purified about 90-fold and shown to be specific for the 7-*O*-position of various flavones, flavonols (but not flavonol 3-*O*-glycosides), and flavanones as acceptors and for UDP-glucose or TDP-glucose as glucosyl donors (Sutter *et al.*, 1972). The substrate specificities (see Table 2) are consistent with the occurrence of both flavone and flavonol 7-*O*-glycosides in the cells used as the source of this enzyme.

More recently, the UDP-glucose:flavonol 3-*O*-glucosyltransferase was also isolated from cell suspension cultures of parsley (Sutter and Grisebach, 1973). This enzyme was completely separated from the 7-*O*-glucosyltransferase. While the 3-*O*-glucosyltransferase is specific for the position of glucosylation, several flavonols, including quercetin 7-*O*-glucoside, can

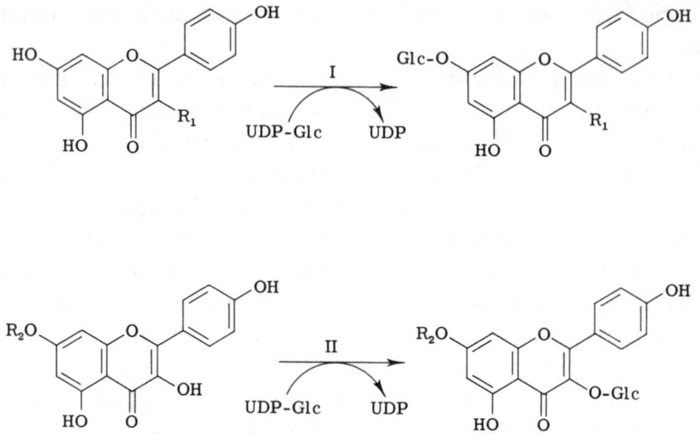

FIG. 13. Reactions catalyzed by two specific glucosyltransferases which were isolated from cell suspension cultures of *Petroselinum hortense*. I, UDP-glucose:flavone/flavonol 7-*O*-glucosyltransferase (R^1 = H or OH); II, UDP-glucose:flavonol 3-*O*-glucosyltransferase (R^2 = H or glucose).

serve as substrates. However, no glucosylation of dihydroquercetin was observed, a finding which supports the conclusion drawn from tracer experiments that glycosylation is the last step in flavonoid glycoside formation (Grisebach, 1965). It seems likely that this enzyme is involved in the formation of both 3-*O*-glucosides and 3,7-di-*O*-glucosides of flavonols in parsley, since no diglucosides are formed by the 7-*O*-glucosyltransferase.

UDP-APIOSE SYNTHASE

In parsley, the branched-chain sugar apiose occurs only in flavone 7-*O*-[β-D-apiofuranosyl (1→2)]-β-D-glucosides (Sandermann, 1969; Kreuzaler and Hahlbrock, 1973). Regulation of its enzymatic formation in young leaves and illuminated cell suspension cultures of this plant is directly related to the biosynthesis of flavone glycosides (Hahlbrock *et al.*, 1971b). The substrate for the apiosyltransferase reaction was identified as UDP-D-apiose (Ortmann *et al.*, 1972) and shown to be formed from UDP-D-glucuronic acid by UDP-apiose synthase (Baron *et al.*, 1973). The enzyme has been isolated and purified from *Lemna minor* (Sandermann and Grisebach, 1970; Wellmann and Grisebach, 1971) and from parsley cell suspension cultures (Wellmann *et al.*, 1971; Baron *et al.*, 1972, 1973).

The reaction mechanism (Fig. 14) involves elimination of the carboxyl group of glucuronic acid as CO_2 and rearrangement of the remaining carbon atoms to form the branched-chain carbon skeleton of apiose (Grisebach

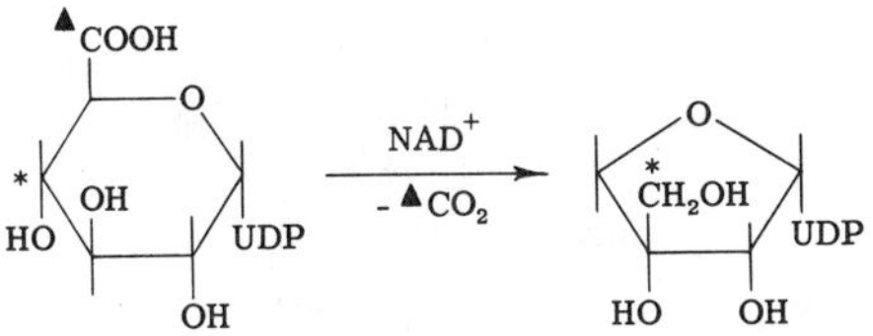

FIG. 14. Formation of UDP-apiose from UDP-glucuronic acid.

and Schmid, 1972). The formation of a second product, UDP-D-xylose, is also catalyzed by the same enzyme. The relative rates of formation of the two sugar nucleotides, UDP-D-apiose and UDP-D-xylose, remained constant throughout a 1000-fold purification of the enzyme. This suggests that the synthase is either an enzyme complex composed of at least two subunits with different catalytic activities or a multifunctional protein with two catalytic sites.

The molecular weight of the UDP-apiose/UDP-xylose synthase is 100,000–115,000 daltons. The enzyme requires NAD+ and thiol reagents for activity. NH4+ inhibits apiose synthesis and stimulates xylose synthesis in the pH range 7.5–8.2 and stimulates both activities at pH 7.0. In these properties it differs markedly from a UDP-xylose synthase also present in parsley cell suspension cultures (Wellmann *et al.*, 1971). While the apiose moiety of UDP-apiose is efficiently transferred to flavone 7-*O*-glucosides, UDP-xylose is not utilized as substrate for the formation of flavonoid glycosides in parsley.

APIOSYLTRANSFERASE

UDP-D-apiose:flavone 7-*O*-[β-D]-glucoside apiosyltransferase catalyzes the formation of flavone 7-*O*[β-D]-apiofuranosyl(1→2)[β-D]-glucosides as shown in Fig. 15 (Ortmann *et al.*, 1970, 1972). The enzyme is highly specific for UDP-D-apiose as glycosyl donor. By contrast, 7-*O*-glucosides of a large variety of flavones, flavanones, and isoflavones, apigenin 7-*O*-glucuronide, and glucosides of *p*-substituted phenols (but not phenolic aglycones) can serve as acceptors. However, no reaction takes place with flavonol 7-*O*-glucosides. This observation is in accordance with the fact that flavonols do not occur in parsley as apiosylglucosides but only as mono- or diglucosides (see above, Structure of Flavonoid Glycosides from Illuminated Cell Suspension Cultures of Parsley).

No cofactors are required for apiosyltransferase activity. The enzyme is rather unstable under various conditions unless purified to some extent and stored in the presence of thiol reagents such as dithioerythritol. Partial purification of the apiosyltransferase resulted in its complete separation

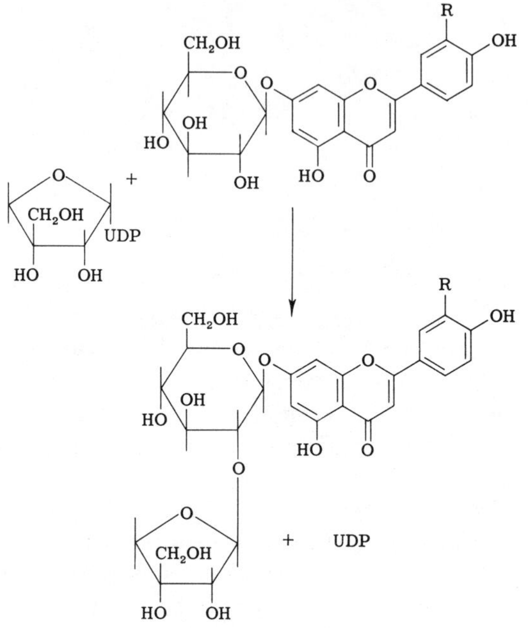

FIG. 15. Formation of apiin (apigenin 7-*O*-[β-D]-apiofuranosyl(1 → 2) [β-D]-glucoside) from apigenin 7-*O*-[β-D]-glucoside and UDP-D-apiose.

from the 7-*O*-glucosyltransferase. The molecular weight of 50,000 daltons is about the same as that determined for the two glucosyltransferases.

All of the data thus far available are consistent with the assumption that flavone apiosylglucosides are formed by a stepwise transfer of glucose and apiose from the corresponding UDP-sugars to the aglycones, as shown in Fig. 6.

MALONYLTRANSFERASE

Malonylation presumably represents the last step of flavonoid glycoside biosynthesis in parsley. The enzyme catalyzing this reaction, malonyl-CoA:flavonoid glycoside malonyltransferase (Fig. 6, enzyme 7), was demonstrated in crude extracts of parsley cell suspension cultures (Hahlbrock, 1972). Malonyl-CoA functions as donor of the acyl residue. Of the number of possible acceptors (cf. Kreuzaler and Hahlbrock, 1973), only apigenin 7-*O*-apiosylglucoside (apiin), and 7-*O*-glucoside (cosmosiin) and chrysoeriol 7-*O*-apiosylglucoside (graveobiosid B) have been tested and shown to be efficiently converted to the corresponding monomalonyl glycosides. Although the exact position of malonylation has not been determined, spectral evidence and the fact that apiin and cosmosiin, but not the aglycone apigenin, served as substrates, suggest that the product of the enzymatic reaction is acylated in the glucose moiety. Furthermore, by analogy with the occurrence of 6″-*O*-malonyl isoflavone glucosides in *Trifolium* species (Beck and Knox, 1971), it can be assumed that the enzyme catalyzes the formation of flavone and flavonol glycosides malonylated in the 6-*O*-position of the glucose.

The malonyltransferase has not been purified to any extent and detailed studies on its substrate specificity and other properties have not been carried out. However, the enzyme was shown to be specifically related to flavonoid biosynthesis by the characteristic changes in its activity after illumination of the cell cultures (Fig. 4).

METHYLTRANSFERASE

As already mentioned (see Section on Hydroxycinnamic Acid 3-*O*-Methyltransferase), it is not yet certain whether *O*-methyl groups are introduced into flavonoids at the phenylpropanoid stage or at a later stage during flavonoid biosynthesis, or both.

Several lines of evidence suggest that a methyltransferase which has been isolated from illuminated cell suspension cultures of parsley is specifically involved in the methylation of flavonoid compounds (Ebel *et al.*, 1972). Thus, the enzyme catalyzes the transfer of the methyl group of *S*-adenosyl-L-methionine (SAM) to 3′,4′-dihydroxyflavonoids exclusively in the *meta* position (Fig. 16). This is in agreement with the occurrence of a number of glycosides of chrysoeriol and isorhamnetin in these cells (Fig 3, Table 1). Furthermore, from the changes in the activity of the SAM: *ortho*-dihydric phenol *meta*-*O*-methyltransferase after illumination of the cell cultures, it is concluded that this methyltransferase is an enzyme of group II, as shown in Fig. 4. Although ferulic acid is also readily formed

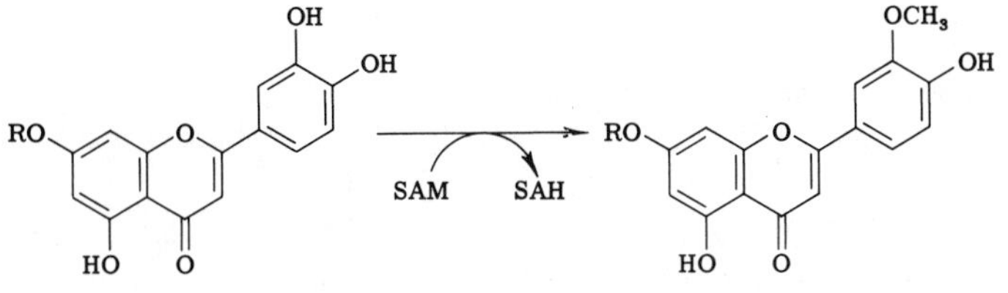

FIG. 16. Enzymatic formation of chrysoeriol from luteolin (R = H) and of chrysoeriol 7-*O*-glucoside from luteolin 7-*O*-glucoside (R = glucose). SAM, *S*-Adenosylmethionine; SAH, *S*-adenosylhomocysteine.

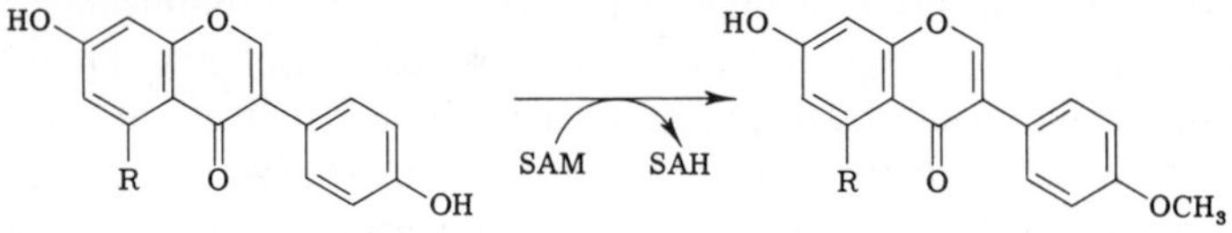

FIG. 17. Enzymatic formation of formononetin from daidzein (R = H) and of bio-chanin A from genistein (R = OH) in cell-free extracts of cell suspension cultures from *Cicer arietinum*. SAM, *S*-Adenosylmethionine; SAH, *S*-adenosylhomocysteine.

from SAM and caffeic acid in the presence of the methyltransferase, the affinity of the enzyme for this acid (K_m = 1.6 $\times$ 10^{-3}) is considerably lower than that for luteolin (K_m = 4.6 $\times$ 10^{-5}) and luteolin 7-*O*-glucoside (K_m = 3.1 $\times$ 10^{-5}).

The partially purified methyltransferase requires Mg^{2+} for optimal en-zymatic activity. In contrast to mammalian catechol *O*-methyltrans-ferases (Molinoff and Axelrod, 1971), the enzyme from parsley cell cultures is not affected by the addition of inhibitors of thiol groups such as *p*-chloro-mercuribenzoate or iodoacetamide. The molecular weight of the methyl-transferase was estimated from the elution volume on a Sephadex G-100 column to be about 48,000 daltons.

Another methyltransferase which also acts at the flavonoid level was found in studies on the biosynthesis of isoflavonoids in cell suspension cultures of *Cicer arietinum* (Wengenmayer, 1972). With a crude enzyme preparation obtained from these cultures, daidzein and genistein were efficiently converted in the presence of *S*-adenosylmethionine to formo-nonetin and biochanin A, respectively (Fig. 17). No reaction was detected with various concentrations of *p*-coumaric or caffeic acid, with the flavones apigenin and luteolin, or with 4,2′,4′-trihydroxychalcone as substrates. This suggests that methylation takes place only at the isoflavone stage, catalyzed by a rather specific *S*-adenosylmethionine:isoflavone 4′-*O*-methyltransferase. Purification of this enzyme is now in progress.

FURTHER ASPECTS OF REGULATION OF FLAVONOID GLYCOSIDE BIOSYNTHESIS

The strictly concomitant changes in activity observed for each of the two groups of enzymes (Fig. 4) suggest highly coordinated mechanisms of regulation of the respective enzyme activities. Although conclusive evidence was not obtained which would prove that these variations in enzyme ac-tivities actually reflect changes in the amounts of enzyme molecules, studies with inhibitors at least showed that synthesis of both RNA and protein is involved in the initial increases in enzyme activities. These experiments further suggested that the lag period of 2.5 hours reflects the period of time required for the sequence of molecular events occurring between RNA synthesis which starts immediately upon illumination and the involvement of protein synthesis in the cytoplasm (K. Hahlbrock, E. Kuhlen, and H. Ragg, unpublished results, 1973). It is tempting to assume that the type of RNA whose synthesis precedes protein synthesis is messenger RNA. Furthermore, it can be speculated that (polycistronic?) heterogeneous nuclear RNAs are precursors of groups of individual messenger RNAs whose translational products would be identical with or related to the enzymes measured. Such a mechanism would be analogous to the recently proposed formation of the two globin messenger RNAs from large heterogeneous nuclear RNAs in globin-synthesizing erythroblasts (Imaizumi *et al.*, 1973).

If this hypothesis is correct, it must be assumed that different rates of synthesis or translation, or both, of messenger RNA and/or different rates of subsequent enzyme inactivation are responsible for the observed differences in the regulation of the enzymes of group I and group II, respectively. One important line of evidence for this assumption was obtained by the observation that the enzymes of group I can be induced independently of those of group II by transferring parsley cells to fresh culture medium (Hahlbrock and Wellmann, 1973). Furthermore, the induction of one or only a few enzymes of either group was never observed without the simultaneous induction of all of the other enzymes of the same group. Hence, it seems likely that the genetic information for each group of enzymes is transcribed simultaneously.

As already mentioned, rapid decreases in enzyme activity following a distinct maximum (Figs. 2 and 4) suggest that not only synthesis but also subsequent inactivation of the enzymes is responsible for the observed changes in activity. While at least some preliminary information concerning the synthesis was obtained by using inhibitors of RNA and protein synthesis, nothing is known as yet with regard to the mechanisms involved in the strikingly coordinated decreases in enzyme activity. Thus, many questions as to the regulation of the enzymes discussed in this chapter will still have to be answered and a number of hypotheses will have to be proven or disproven. The very fact, however, that the enzymes of groups I and II in parsley cell cultures as well as at least some of the enzymes of group I in soybean cell cultures (cf. Hahlbrock *et al.*, 1971c) are regulated in such a highly coordinated manner, suggests that these groups of enzymes provide an excellent system for further studies on the molecular mechanisms related to the simultaneous expression of enzyme activities in eukaryotic cells.

Enzymology of Lignin Biosynthesis

FORMATION OF CONIFERYL ALCOHOL

Tracer studies and model experiments have amply demonstrated that cinnamyl alcohols are the primary building stones of lignin (Freudenberg and Neish, 1968; Sarkanen and Ludwig, 1971). Evidence for the reduction of ferulic acid to coniferyl alcohol via coniferyl aldehyde has been obtained from tracer experiments (Higuchi and Brown, 1963). The reduction of the phenylpropanoid acids to the alcohols would be expected to require activation of the carboxyl group, most likely by formation of the coenzyme A esters. The formation of coenzyme A thiol esters of cinnamic acids has already been discussed (*p*-Coumarate:CoA Ligase, above). Recently an enzyme system catalyzing the reduction of feruloyl-CoA to coniferyl alcohol (Fig. 18) has been found in our laboratory and by Zenk and associates.

As the source of the enzyme we have used cell suspension cultures of soybean (*Glycine max*). These cultures form a ligninlike substance during their growth. Lignin was isolated according to the Klason procedure (Carceller *et al.*, 1971) and characterized by its color reaction with phloroglucinol-HCl and by nitrobenzene oxidation (Whitmore, 1971) which gave two compounds which corresponded to vanillin and *p*-hydroxybenzaldehyde (J. Ebel and H. Grisebach, unpublished results, 1973) in R_f values on silica gel plates.

When crude extracts of the cells or extracts treated with Dowex I acetate were incubated with [2-^{14}C]-ferulic acid, ATP, CoASH, Mg^{2+}, and NADPH, only one radioactive product was formed, which proved to be identical with coniferyl alcohol (Ebel and Grisebach, 1973). With free ferulic acid as substrate, the reduction requires ATP, Mg^{2+}, CoASH, and NADPH. NADH has about 10 percent of the activity of NADPH. When the CoA

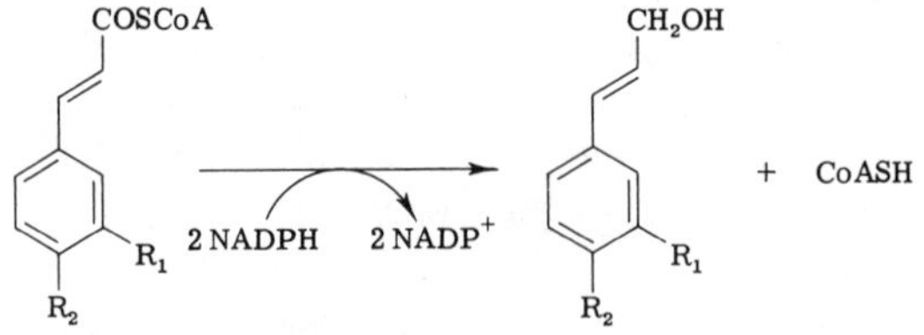

(I) R$_1$ = H ,   R$_2$ = OH
(II) R$_1$ = OCH$_3$, R$_2$ = OH
(III) R$_1$ = OCH$_3$, R$_2$ = OCH$_3$

FIG. 18. Reduction of cinnamoyl-CoA esters with an enzyme system from cell suspension cultures of soybean.

ester is used as substrate, the only cofactor required for the production of coniferyl alcohols is NADPH (J. Ebel and H. Grisebach, unpublished results, 1973). *p*-Coumaroyl-CoA is reduced by the same enzyme preparation to *p*-coumaryl alcohol.

Recently we were able to show that the reduction proceeds in two steps via the aldehyde as suggested by tracer experiments (Higuchi and Brown, 1963):

1. Feruloyl-CoA + NADPH $\rightleftharpoons$ Coniferylaldehyde + NADP$^+$ + H$^+$

2. Coniferylaldehyde + NADPH $\rightleftharpoons$ Coniferylalcohol + NADP$^+$ + H$^+$

The cinnamoyl-CoA reductase and the aromatic aldehyde reductase could be separated on a DEAE cellulose column (J. Ebel and H. Grisebach, unpublished results). Whether the first reaction is reversible and if it proceeds via an enzyme-bound aldehyde as in the case of 3-hydroxy-3-methylglutaryl-coenzyme A reductase (Rétey *et al.*, 1970) is an open question. The aromatic aldehyde reductase was separated on the DEAE column into two isoenzymes and could also be separated from an aliphatic alcohol dehydrogenase (D. Wyrambik and H. Grisebach, unpublished results).

Mansell *et al.* (1972) reported on a cell-free system from cambial tissue of *Salix alba* that catalyzed the reduction of ferulic acid to coniferyl alcohol in the presence of ATP, coenzyme A, and reduced pyridine nucleotides. In a paper from the same laboratory it was later reported that, with an enzyme preparation from young stems of *Forsythia sp.*, it was possible to resolve the reduction of feruloyl-CoA to coniferyl alcohol into two reactions (Gross *et al.*, 1973). The coniferyl aldehyde formed from feruloyl-CoA in the presence of NADPH was trapped as the phenylhydrazone. Enzymatic activity for the reduction of the aldehyde to coniferyl alcohol could also be demonstrated to be present in the crude extracts. Details on the properties of these enzymes are as yet unknown.

The cinnamic acids are not only precursors of lignin but also of other phenylpropanoid derivatives (Fig. 5). The reduction of the acids to the corresponding alcohols, which are the primary building stones of lignin, is therefore an important step directing the phenylpropanoid pathway toward lignin biosynthesis. It can therefore be expected that enzymes involved in such a step are under regulatory control. Further work on the enzymes mentioned above can therefore give important clues to the regulation of lignification.

POLYMERIZATION OF CINNAMYL ALCOHOLS TO LIGNIN

The two enzymes laccase (*p*-diphenol:O$_2$ oxidoreductase) and peroxidase (donor:H$_2$O$_2$ oxidoreductase) have been implicated in the polymerization

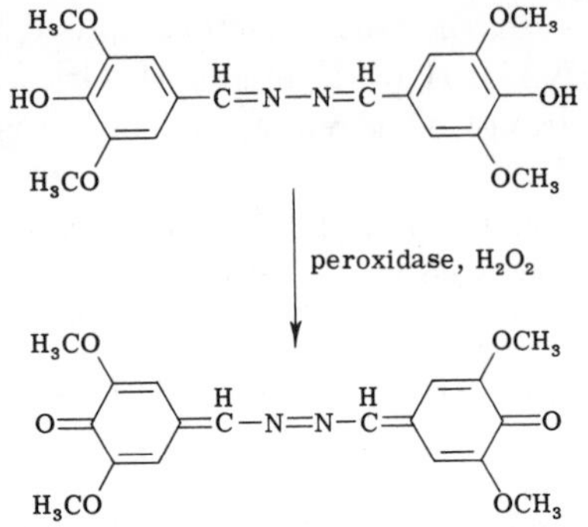

FIG. 19. Oxidation of syringaldazine by peroxidase and H_2O_2 to the purple tetramethoxyazo-p-methylenequinone.

of p-hydroxycinnamyl, coniferyl, and sinapyl alcohol to lignin (Freudenberg and Neish, 1968; Sarkanen and Ludwig, 1971). The participation of laccase was mainly deduced from the experiments of the Freudenberg group using a fungal laccase to make biosynthetic lignins (dehydrogenation polymer) *in vitro* from cinnamyl alcohols. In a recent investigation by Harkin and Obst (1973), the participation of the two enzymes mentioned above in lignin formation was estimated by application of a histochemical method. Syringaldazine is oxidized in the presence of laccase and atmospheric oxygen to the intensely purple tetramethoxyazo-p-methylenequinone (Fig. 19). Peroxidase plus H_2O_2 produces the same effect.

Application of one or two drops of an 0.1 percent solution of syringaldazine in ethanol to cross sections of freshly microtomed surfaces of numerous sample stubs from angiosperms and gymnosperms did not produce even the faintest coloration to indicate the presence of laccase. However, when one or two drops of an 0.03 percent aqueous H_2O_2 were added to the section or end surface, an intense purple ring formed almost immediately in the xylem tissue adjacent to the cambium. From these and further results not described here the authors conclude that it therefore seems certain that the phenol oxidase in the zone of incipient lignification is exclusively peroxidase. If this conclusion can be confirmed with other methods, the interesting question arises as to the nature of the reactions which generate all the H_2O_2 needed for lignification. More information is also needed on the peroxidases which are present in lignifying tissues.

ACKNOWLEDGMENT

Our own results discussed in this review would not have been possible without the hard work of a number of very able and enthusiastic co-workers whose names are mentioned in the appropriate references. Our work was supported by Deutsche Forschungsgemeinschaft (SFB 46) and by Fonds der Chemischen Industrie.

REFERENCES

Baron, D., E. Wellmann, and H. Grisebach. 1972. *Biochim. Biophys. Acta* **258**:310–318.

Baron, D., U. Streitberger, and H. Grisebach. 1973. *Biochim. Biophys. Acta* **293**:526–533.

Beck, A. B., and J. R. Knox. 1971. *Aust. J. Chem.* **24**:1509–1518.

Birch, A. J., and F. W. Donovan. 1953. *Aust. J. Chem.* **6**:360–368.

Büche, T. 1973. Staatsexamensarbeit, University Freiburg/Br., Germany.

Büche, T., and H. Sandermann, Jr. 1973. *Arch. Biochem. Biophys.* **158**:445–447.

Camm, E. L., and G. H. N. Towers. 1973. *Phytochemistry* **12**:961–973.

Carceller, M., M. R. Davey, M. W. Fowler, and H. E. Street. 1971. *Protoplasma* **73**:367–385.

Dimroth, P., H. Walter, and F. Lynen. 1970. *Eur. J. Biochem.* **13**:98–110.

Ebel, J., and H. Grisebach. 1973. *FEBS Lett.* **30**:141–143.

Ebel, J., K. Hahlbrock, and H. Grisebach. 1972. *Biochim. Biophys. Acta* **269**:313–326.

Endress, R. 1972. *Z. Pflanzenphysiol.* **67**:188–191.

Finkle, B. J., and M. S. Masry. 1964. *Biochim. Biophys. Acta* **85**:167–169.

Finkle, B. J., and R. F. Nelson. 1963. *Biochim. Biophys. Acta* **78**:747–749.

Freudenberg, K., and A. C. Neish. 1968. "Constitution and Biosynthesis of Lignin." Springer-Verlag, Berlin and New York.

Grambow, H. J., and H. Grisebach. 1971. *Phytochemistry* **10**:789–796.

Grisebach, H. 1962. *Planta Med.* **10**:385–397.

Grisebach, H. 1965. *In* "Chemistry and Biochemistry of Plant Pigments" (T. W. Goodwin, ed.), pp. 279–308. Academic Press, New York.

Grisebach, H. (1967). *In* "Biosynthetic Patterns in Microorganisms and Higher Plants," pp. 1–31. Wiley, New York.

Grisebach, H. 1968. *Recent Advan. Phytochem.* **1**:379–406.

Grisebach, H., and W. Barz. 1969. *Naturwissenschaften* **56**:538–544.

Grisebach, H., and G. Brandner. 1961. *Z. Naturforsch.* **B16**:2–5.

Grisebach, H., and G. Brandner. 1962. *Biochim. Biophys. Acta* **60**:51–57.

Grisebach, H., and U. Döbereiner. 1964. *Biochem. Biophys. Res. Commun.* **17**:737–741.

Grisebach, H., and R. Schmid. 1972. *Angew. Chem.* **84**:192–206; *Angew. Chem., Int. Ed. Engl.* **11**:159–173 (1972).

Grisebach, H., W. Barz, K. Hahlbrock, S. Kellner, and L. Patschke. 1966. *In* "Biosynthesis of Aromatic Compounds" (G. Billek, ed.), pp. 25–36. Pergamon, Oxford.

Gross, G. G., J. Stöckigt, R. L. Mansell, and M. H. Zenk. 1973. *FEBS Lett.* **31**:283–286.

Hahlbrock, K. 1972. *FEBS Lett.* **28**:65–68.

Hahlbrock, K., and H. Grisebach. 1970. *FEBS Lett.* **11**:62–64.

Hahlbrock, K., and H. Grisebach. 1973. *In* "The Flavonoids" (T. J. Mabry, and J. B. Harborne, eds.). Chapman & Hall, London (in press).

Hahlbrock, K., and E. Kuhlen. 1972. *Planta* **108**:271–278.

Hahlbrock, K., and E. Wellmann. 1970. *Planta* **94**:236–239.

Hahlbrock, K., and E. Wellmann. 1973. *Biochim. Biophys. Acta* **304**:702–706.

Hahlbrock, K., E. Wong, L. Schill, and H. Grisebach. 1970a. *Phytochemistry* **9**:949–958.

Hahlbrock, K., H. Zilg, and H. Grisebach. 1970b. *Eur. J. Biochem.* **15**:13–18.

Hahlbrock, K., A. Sutter, E. Wellmann, R. Ortmann, and H. Griesbach. 1971a. *Phytochemistry* **10**:109–116.

Hahlbrock, K., J. Ebel, R. Ortmann, A. Sutter, E. Wellmann, and H. Grisebach. 1971b. *Biochim. Biophys. Acta* **244**:7–15.

Hahlbrock, K., E. Kuhlen, and T. Lindl. 1971c. *Planta* **99**:311–318.

Hanson, K. R., and E. A. Havir. 1972a. *In* "The Enzymes" (P. D. Boyer, ed.), 3rd ed., Vol. 7, pp. 75–166. Academic Press, New York.

Hanson, K. R., and E. A. Havir. 1972b. *In Recent Advan. Phytochem.* **4**:45–85.

Harkin, J. M., and J. R. Obst. 1973. *Science* **180**:296–298.

Havir, E. A., and K. R. Hanson. 1973. *Biochemistry* **12**:1583–1591.

Hess, D. 1966. *Z. Pflanzenphysiol.* **55**:374–386.

Higuchi, T., and S. A. Brown. 1963. *Can. J. Biochem. Physiol.* **41**:621–628.

Imaizumi, T., H. Diggelmann, and K. Scherrer. 1973. *Proc. Nat. Acad. Sci. U.S.* **70**:1122–1126.

Knobloch, K. H., and K. Hahlbrock, (1973). *Hoppe-Seyler's Z. Physiol. Chem.* **354**:1210.

Koukol, J., and E. Conn. 1961. *J. Biol. Chem.* **236**:2692–2698.

Kreuzaler, F., and K. Hahlbrock. 1972. *FEBS Lett.* **28**:69–72.

Kreuzaler, F., and K. Hahlbrock. 1973. *Phytochemistry* **12**:1149–1152.

Lindl, T., F. Kreuzaler, and K. Hahlbrock. 1973. *Biochim. Biophys. Acta* **302**:457–464.

Mansell, R. L., and J. A. Seder. 1971. *Phytochemistry;* **10**:2043–2045.

Mansell, R. L., J. Stöckigt, and M. H. Zenk. 1972. *Z. Pflanzenphysiol.* **68**:286–288.

Molinoff, P. B., and J. Axelrod. 1971. *Annu. Rev. Biochem.* **40**:465–500.

Neish, A. C. 1961. *Phytochemistry* **1**:1–24.

Neiph, A. C. 1964. *In* "Biochemistry of Phenolic Compounds" (J. B. Harborne, ed.), pp. 295–360. Academic Press, New York.

Neujahr, H. Y., and A. Gaal. 1973. *Eur. J. Biochem.* **35**:386–400.

Ortmann, R., H. Sandermann, Jr., and H. Grisebach. 1970. *FEBS Lett.* **7**:164–166.

Ortmann, R., A. Sutter, and H. Grisebach. 1972. *Biochim. Biophys. Acta* **289**:293–302.

Pachéco, H. 1969. *Bull. Soc. Frn. Physiol. Veg.* **15**:3–28.

Rétey, J., E. von Stetten, U. Coy, and F. Lynen. 1970. *Eur. J. Biochem.* **15**:72–76.

Roberts, R. J., and P. F. T. Vaughan. 1971. *Phytochemistry* **10**:2649–2652.

Russell, D. W. 1971. *J. Biol. Chem.* **246**:3870–3878.

Russell, D. W., and E. E. Conn. 1967. *Arch. Biochem. Biophys.* **122**:256–258.

Sandermann, H., Jr. 1969. *Phytochemistry;* **8**:1571–1575.

Sandermann, H., Jr., and H. Grisebach. 1970. *Biochim. Biophys. Acta* **208**:173–180.

Sarkanen, K. V., and C. H. Ludwig. 1971. "Lignins." Wiley (Interscience), New York.

Schill, L., and H. Grisebach. 1973. *Hoppe-Seyler's Z. Physiol. Chem.* **354**:1555–1562.

Shimada, M., H. Ohashi, and T. Higushi. 1970. *Phytochemistry* **9**:2463–2470.

Shimada, M., H. Fushiki, and T. Higushi. 1972. *Phytochemistry* **11**:2657–2662.

Sutter, A., and H. Grisebach. 1973. *Biochim. Biophys. Acta* **309**:289–295.

Sutter, A., R. Ortmann, and H. Grisebach. 1972. *Biochim. Biophys. Acta* **258**:71–87.

Vaughan, P. F. T., and V. S. Butt. 1969. *Biochem. J.* **113**:109–115.

Vaughan, P. F. T., and V. S. Butt. 1970. *Biochem. J.* **119**:89–94.

Vaughan, P. F. T., V. S. Butt, H. Grisebach, and L. Schill. 1969. *Phytochemistry* **8**:1373–1378.

Wellmann, E., and H. Grisebach. 1971. *Biochim. Biophys. Acta* **235**:389–397.

Wellmann, E., D. Baron, and H. Grisebach. 1971. *Biochim. Biophys. Acta* **244**:1–6.

Wengenmayer, H. 1972. Diplomarbeit, Freiburg/Br., Germany.

Whitmore, F. W. 1971. *Plant Physiol.* **48**:596–602.

Wong, E., and E. Moustafa. 1966. *Tetrahedron Lett.* pp. 3021–3022.

POSSIBLE MULTIENZYME COMPLEXES REGULATING THE FORMATION OF C$_6$-C$_3$ PHENOLIC COMPOUNDS AND LIGNINS IN HIGHER PLANTS

H. A. STAFFORD

Biology Department, Reed College,
Portland, Oregon

Introduction

Tissues of higher plants accumulate a wide variety of phenolic compounds such as the esters of C$_6$–C$_3$ phenolic compounds, polymers of C$_6$–C$_3$

units or lignins, and C_{15} flavonoids (Fig. 1). All are dependent on the shikimic acid pathway producing phenylalanine and tyrosine. Since these products may be formed within the same cell, and frequently are accumulated independently in vacuoles or in cell walls, an understanding of the regulation of these pathways is of prime importance, as stressed by Creasy and Zucker in this volume.

Since most of the intermediates involved in the initial C_6–C_3 hydroxylation sequence are also substrates for a variety of subsequent enzymatic reactions, it is unlikely that the enzymes or their substrates are randomly distributed throughout the cytosol. In fact, most "soluble" enzymes may be at least loosely associated with membrane systems (Ginsburg and Stadtman, 1970). Furthermore, enzymes that operate in a biosynthetic sequence might be expected to be found in multienzyme complexes, preferably attached to cytoplasmic membrane surfaces or enclosed within organelles.

The sequence of six chemical reactions converting phenylalanine to sinapic acid can be used to illustrate the concept of multienzyme complexes and the regulation of their enzymatic activities (Fig. 1). These reactions involve three hydroxylations and two methylations following the initial

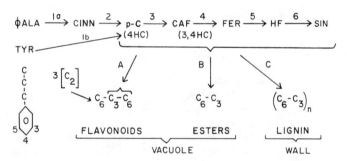

FIG. 1. C_6–C_3 sequence leading to flavonoids, esters, and lignin. Steps 1–6 may be catalyzed by five enzymes, if one ammonia-lyase acts on both phenylalanine and tyrosine, and if one methyl transferase acts in steps 4 and 6. These five enzymes are listed in Table 1, with cofactors for the reaction. Pathway A involves the activation of the carboxyl group of *p*-coumarate and subsequent condensation with $3[C_2]$ units via malonyl-CoA to form flavonoids (see Grisebach, this volume). Pathway B leads to glucose esters of the various C_6–C_3 acids via UDP-glucose (see Stafford, 1974, for the special case of quinic esters of caffeic). In Pathway C, the carboxyls of *p*-coumarate, ferulic, and sinapic acids are activated via CoA ligases, and the resulting CoA esters are reduced to aldehydes and alcohols which are subsequently oxidatively polymerized to lignin in the cell wall by peroxidase. φAla, phenylalanine; tyr, tyrosine; cinn, *trans*-cinnamic; p-C, *p*-coumaric or 4HC, 4-hydroxycinnamic; CAF, caffeic or 3,4HC, 3,4-dihydroxycinnamic; FER, ferulic; HF, 5-hydroxyferulic; SIN, sinapic acids.

TABLE 1

ENERGYMES IN THE C_6–C_3 BIOSYNTHETIC SEQUENCE

	Abbreviation	Cofactors
1. Phenylalanine and tyrosine ammonia-lyases	Ly	—
2. Cinnamate 4-hydroxylase (Cyt-P_{450})[a]	CH	NADPH
3. 4-Hydroxycinnamate 3-hydroxylase (*p*-coumaric hydroxylase)[a]	HCH	Ascorbate or NADPH or reduced pteridines
4. *O*-Methyltransferase	MT	Mg^{2+}, S-AME
5. Ferulate 5-hydroxylase[a]	FH	?

[a] Probably all are monooxygenases (mixed-function oxidases).

ammonia-lyase step (Table 1). While the compounds in this sequence are involved in other reactions than those shown, I will concentrate on the formation of the C_6–C_3 esters and lignin. Other reactions have been reviewed elsewhere (Stafford, 1974), and several are described in other chapters of this volume.

Some of the basic problems in the understanding of the regulation of such a biosynthetic sequence are as follows:

(1) Some of the above phenolic compounds in the sequence leading to sinapic acid are both *intermediates* and *final products* which can be accumulated as esters or are used in other pathways such as lignin and flavonoid biosyntheses.

(2) The products accumulated are found in the large central vacuole or in the cell wall.

(3) All of the three types of products (flavonoids, esters, lignin) may be found at one time in the same cell.

(4) Specific compounds with the same basic hydroxylation pattern may be accumulated independently.

Multienzyme complexes found in the cytosol ("soluble") fractions have been recently discussed (Ginsburg and Stadtman, 1970). One of these of particular interest here as a prototype is the aggregate found in *Neurospora* which is involved in the shikimate pathway. The *arom* cluster encodes five enzymes which catalyze five steps leading ultimately to chorismate, after the initial condensation of erythrose-4-P + PEP (Fig. 2). This aggregate (molecular weight of about 230,000) consists of two identical subunits, each containing the five activities believed to be controlled by four to five unique polypeptides. Two of the activities which always remain tightly associated in *Neurospora* are also found together in some algae and higher

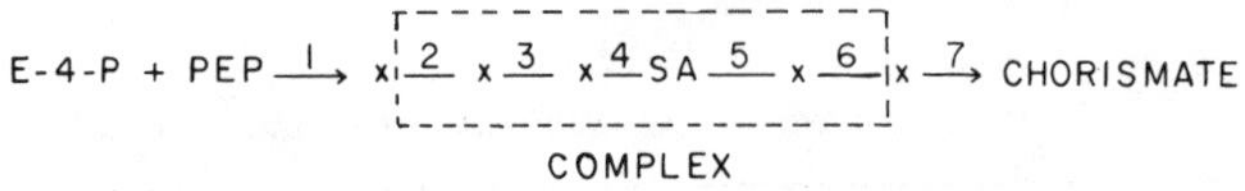

$$E\text{-}4\text{-}P + PEP \xrightarrow{\;1\;} x\underline{\;2\;} x\underline{\;3\;} x\underline{\;4\;}SA\underline{\;5\;} x\underline{\;6\;}x \xrightarrow{\;7\;} CHORISMATE$$

COMPLEX

Fig. 2. A multienzyme complex in *Neurospora* coded by the *arom* cluster of genes, catalyzing steps 2 through 6 in the pathway to chorismate from erythrose-4-P (E-4-P) + P-enol pyruvate (PEP).

plants, but in these cases are not associated with a large aggregate. The complex may be dissociated or broken down by protease activity during extraction in these tissues. The aggregate in *Neurospora* preferentially utilizes precursors formed in the biosynthetic sequence rather than those intermediates supplied externally (catalytic facilitation) (Gaertner *et al.*, 1970; Jacobson *et al.*, 1972). Other multienzyme complexes related to phenolic metabolism are acetyl-CoA carboxylase and fatty acid synthetase, which also are found as separate or dissociated enzymes in some tissues (Sedgwick, 1973; Sumper and Lynen, 1972). Multienzyme complexes found in organelles such as mitochondria are well known.

Sorghum as a Biological Test Organism

TISSUE TYPES USED

Two types of tissues of *Sorghum*, a C_4 type monocot, have been used in my studies—the first internode (mesocotyl) and the leaf (including sheath plus lamina or blade). The basic structures of this monocot are diagrammed in Fig. 3, with comparisons with the hypocotyl of the dicot counterpart used in many other studies of phenolic metabolism.

The internal anatomy of the *first* internode of *Sorghum* [*Sorghum bicolor* (Linn) Moench., formerly called *S. vulgare* (Doggett, 1970)] is similar to that of *Zea mays*. Both differ slightly from *Avena* and *Phleum*, which have a cortical vascular bundle in addition to the central stele (Avery, 1930). In all of these, the *first* internode is light-sensitive, growing beyond a microscopic size only in the dark (Fig. 3). Both groups differ considerably in seedling anatomy from wheat and barley, in which the *second* internode is light-sensitive, the result being that the structure growing in the dark includes second internode, coleoptile, and internal young leaves, making it a very complex seedling tissue to interpret. These differences result from differing positions of the meristem in the embryo. Note that the meristem in the first internode is apical, so that a gradient of differentiation exists with the older tissue toward the base. Subsequent internodes, however, have the typical intercalary meristem of grasses found at the base of the internodes. There is a sequential growth of these seedling structures, the first internode elongating first, followed by that of the coleoptile after the growth of the first internode begins to stop. Upon excision, the first internode ceases growth unless exogenous hormones are added, so that subsequent changes in the internode are not the result of growth. These excised internodes can be kept in a moist environment for 3 to 6 days, during which time the enzyme levels change and there is a large increase in phenolic compounds using endogenous starch reserves. Greening of tissues rarely occurs in either first internode or coleoptilar tissue. [Proplastids are present, and in some cases in coleoptilar tissues they turn slightly green (Hinchman, 1972).] The basic anatomy of the first internode is more rootlike than stemlike, consisting of a central solid stele of vascular tissue surrounded by the endodermal layer and the cortex which makes up 90 percent of the volume of a root. The stele and cortex can be easily separated.

Since hypocotyl tissue found in some dicots has been widely studied (Schopfer and Mohr, 1972; Mohr, 1972; Engelsma, 1972), a brief comparison with this structure in a typical dicot is useful (Fig. 3). The growth of both tissues is inhibited by light. Both have similarly oriented differentiation gradients with older tissue at the base. They differ, however, in the position of the source of the stored starch prior to excision; the supply is

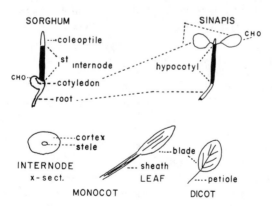

Fig. 3. Seedling and leaf characteristics of *Sorghum*, a monocot, and *Sinapis*, a dicot. Due to differences in position of the growing portions of the shoots relative to the cotyledons, the epicotyl in a grass such as *Sorghum* consists of 1st internode (or mesocotyl) plus the coleoptile containing the enclosed embryonic growing point and young leaves, while that of a dicot such as *Sinapis* consists only of the embryonic growing point and leaves above the point of attachment of the two cotyledons. Note that since the single cotyledon of the monocot is attached to the base of the first internode, the term "mesocotyl" is inappropriate.

basal in the monocot first internode, since the endosperm and cotyledon in the seed are attached basally, while the supply is apical in dicots, coming from the pair of cotyledons attached to the top of the hypocotyl. Published articles are not always clear as to what portion of the shoot is being used as an enzyme source and its treatment prior to excision. For instance, the light treatments that I will discuss are given to the first internode after excision, and only the first internode was extracted. In studies with the hypocotyl of *Gherkin* and *Sinapis*, the light treatments are frequently given to the intact shoot (hypocotyl plus cotyledons) and the entire shoot is sometimes subsequently extracted so that a mixture of shoot tissues is involved. Furthermore, if the hypocotyl or first internode is left intact during treatment, subsequent growth is still possible unless light intensities are quite high. Another dicot frequently used in studies of phenolic metabolism is the pea plant; its seedling growth and development differ from the typical dicots in the absence of a hypocotyl (see Mohr, 1972, for diagram).

The leaf of a dicot differs considerably from that of a monocot in shape and the amount and arrangement of vascular tissue, but the major difference for the purposes of this discussion is that the leaves of a dark-grown monocot continue to develop normally except for the presence of etioplasts rather than chloroplasts. Stem growth, on the other hand, is quite limited in the dark. In contrast, leaves of a dicot seedling do not develop beyond embryonic stages in the dark, with most of the growth made up of stem tissues. It should also be remembered that *Sorghum* leaves have the type of anatomy associated with the C_4 photosynthetic pathway.

PHENOLIC COMPOUNDS IN *Sorghum*

The major phenolic compounds identified so far in *Sorghum* seedlings and young green leaves are summarized in Table 2. I wish to emphasize the following:

(1) The mono-, di-, and trihydroxylation patterns are varied, with the trihydroxy form being found only in the ester and lignin group.

(2) Large amounts of the diphenol, caffeic acid, are present in green leaves with only trace amounts in etiolated leaves or seedling tissues.

(3) The hydroxylation pattern of the B-ring of flavonoids varies independently of that of the C_6–C_3 esters as well as within the anthocyanin group. Both HIR and P_{fr} are obligatory requirements only for the cyanidin type (Downs and Siegelman, 1963; Stafford, 1965, 1966). Luteolinidin with a similar B-ring is produced in the dark along with apigeninidin in first internodes. Interactions of the light effect with ethylene have recently been reported (Craker and Wetherbee, 1973). Luteolin, on the other hand,

TABLE 2

PHENOLIC COMPOUNDS IN *Sorghum*

Compound	Structure class	1st Internode	Leaf	+OH	+CH₃
Dhurrin	C₆–C₂ (CN-Glycoside)	+	Tr-young	4	0
Apigeninidin	C₁₅ Glycosides Flavonoids (B-Ring)	+	–	4'	0
Luteolinidin		+		3',4'	0
Cyanidin		+[a]	+[b]	3',4'	0
Luteolin		–	+[b]	3',4'	0
4-OH-Cinnamic		+	+	4	0
Caffeic	C₆–C₃ Esters (glucose ?)	Tr[b]	+[b]	3,4	0
Ferulic		+	++	3,4	3
Sinapic		++	++	3,4,5	3,5
Lignin	(C₆–C₃)ₙ	+	+	4	0
				3,4	3
				3,4,5	3,5

Tissue groupings: Cortex · Vascular tissue — Vacuole / Wall

[a] HIR + P_FR (Downs and Siegelman, 1963).

[b] Light-dependent. Action spectrum unknown. (Stafford, 1969; Stafford and Bliss, 1973.)

may require high-intensity light as it is either in very low concentration or is not detectable in etiolated leaves. Its control may be similar to that of caffeic acid, both of these having a diphenol pattern of hydroxylation. Therefore, while a light requirement may be obligatory for the 3,4-hydroxylation pattern in leaves, it varies with the compound in first internodes and may be controlled by different genetic factors (Stafford, 1970).

(4) The flavonoids probably are found only in the cortex, while lignin, with all three basic hydroxylation patterns, is found only in the stelar tissues of *Sorghum*. Except possibly for sinapic acid, all C_6–C_3 esters are found in both areas and appear to vary independently (Stafford, 1969).

(5) All are accumulated in the vacuoles, except for lignin, which is a wall product. The regulation of the synthesis and accumulation of the various hydroxylation patterns is obviously not a simple matter and requires considerable control.

Potential Enzyme Complexes in the C_6–C_3 Sequence

Assuming that the C_6–C_3 conversions exist *in vivo* as a multienzyme complex, is there only one complex with a minimum of five enzymes or might there be a separate complex for each of the hydroxylated cinnamic acid derivatives that is accumulated? If there is only one multienzyme complex, the regulatory control must involve agents which affect the relative rates of different activities of the complex in order to permit the accumulation of different amounts of the various intermediates (as products). This single type of complex could be randomly distributed in the cytosol, associated with one specific intracellular site, or at least temporarily bound to different sites via different binding groups. But this same complex would have to serve in the accumulation of esters and flavonoids in vacuoles and of lignin in cell walls, unless separated in different cells or active at different times in the development of the cell.

If, on the other hand, there is a separate type of complex for each of the C_6–C_3 end products, there could be populations of different-sized complexes, and these could be randomly distributed or attached to different sites, as in the case of a single complex, giving rise to a series of isozymes when separated from the complexes. Such a series of different complexes would have the advantage of control at the level of the gene and the ribosome as well as at the level of the activity of the individual complexes. These two possibilities are summarized in Fig. 4. The four complexes shown represent the assemblages of enzymes leading to the four hydroxycinnamic acids which accumulate in *Sorghum*, i.e., 4-hydroxycinnamic, caffeic,

```
        ENZYMES              PRODUCTS

 I.  Ly-CH                p-COUMARIC              COOH
                             (4HC)                 CH
 2.  Ly-CH-HCH             CAFFEIC                  CH

 3.  Ly-CH-HCH-MT          FERULIC
                                                    O
 4.  Ly-CH-HCH-MT          SINAPIC          5         3
                  FH                                 4
```

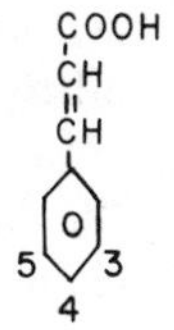

FIG. 4. Hypothetical complexes capable of accumulating four different phenolic end products. Ly, phenylalanine ammonia-lyase; CH, cinnamate 4-hydroxylase; HCH, 4-hydroxycinnamate 3-hydroxylase; MT, *O*-methyl transferase; FH, ferulate 5-hydroxylase.

ferulic, and sinapic acids.* If there is only one type of complex, it should, therefore, be type 4.

Three predictions might be made based on such complexes: one is that 5-hydroxyferulic acid would not accumulate because the methyltransferase necessary to produce ferulic acid would also methylate its hydroxylated product, since transmethylases of angiosperms are able to methylate either hydroxyl group in the 3- or 5-positions (Shimada *et al.*, 1972). Although 5-hydroxyferulic acid has been shown to be a precursor of sinapic and syringyl units, it does not accumulate in plant tissues (Higuchi, 1971; Shimada *et al.*, 1972). The above prediction could be true regardless of the number of complexes, but would be unavoidable if type 4 complexes occur.

The second prediction is that if there were different complexes, the one producing sinapic acid would be the least abundant since it may be limited to the stelar tissue of *Sorghum*. Ferulate-5-hydroxylase activity appears only in this fourth type. The low activity may explain why this third hydroxylation step has not been demonstrated in a cell-free system. In contrast, the combined lyase and cinnamate hydroxylase activities should be the highest.

The third prediction is that since green leaves accumulate all four C_6–C_3 phenolic products, all four complexes could be present, while internode tissues that accumulate all but caffeic could have only three. The presence of large amounts of flavonoids with a diphenol B-ring could be due to hydroxylation at the C_{15} level (Harborne, 1973).

---

* The majority of workers use the trivial name *p*-coumaric acid for 4-hydroxycinnamic acid, but the latter is more informative when the enzymes involved are discussed. I will use the two terms interchangeably for the substrate, but will refer to the hydroxylase as 4-hydroxycinnamate hydroxylase (HCH).

In the schemes in Fig. 4, I have assumed that phenylalanine is the initial substrate via PAL activity. If tyrosine is the substrate via TAL activity, the cinnamic acid hydroxylase step would be by-passed. If there are separate lyases for tyrosine and phenylalanine, a greater number of potential complexes may be involved.

I have not included in any complex the glucosylation step via UDP-glucose to form the glucose esters [see Zenk and Gross (1972) for a discussion of this step as one example of activation of the carboxyl group]. As indicated in the scheme in Fig. 1, I prefer to consider this glucosylation as a terminal step, possibly related to the accumulation of these products in the central vacuole.

In the following section, I will emphasize some recent experiments concerned with the association of 4-hydroxycinnamate hydroxylase (HCH) activity with a minimum of three different molecular weight forms. I will speculate that they may be parts of multienzyme complexes, and that one might expect HCH activity to be associated with three types of complexes leading to caffeic, ferulic, and sinapic acids, respectively.

Enzymes of the C_6–C_3 Sequence

Enzymological Data from *Sorghum*

The Overall Sequence

In experiments reported previously (Stafford and Baldy, 1970), small amounts of ^{14}C-ferulic acid were obtained when ^{14}C-phenylalanine, -tyrosine, or -cinnamate were added to a crude particulate fraction from first internodes of *Sorghum* in the presence of appropriate cofactors such as NADPH, s-AME (Table 1), and ascorbic acid (to block an interfering monophenol oxidase reaction catalyzed by peroxidase if mercaptoethanol is present.) But the yields were low: only 1 percent starting with ^{14}C-labeled cinnamic acid. Little radioactivity was found in the area of caffeic acid, an expected result if multienzyme complexes are involved.

All of the enzymes in the sequence up to the level of ferulic acid could be assayed for independently in the particulate fractions, although some were found also in the "soluble" fraction (Stafford, 1969; Stafford and Dresler, 1972). Unlike many dicots, in which Zucker (1972) postulated a light-induced PAL operon, no photoinduction of lyase activity could be detected in first internodes. This lack of photoinduction may be characteristic of monocots with light-sensitive first internodes and could represent a difference in control mechanisms. McClure (1974) has reported a light induction of both PAL and TAL in etiolated shoots of barley. Since this monocot has a light-sensitive second internode, the tissue extracted consisted of internodal, coleoptilar, and inner leaf tissue.

Cinnamate hydroxylase was very active in 3- to 4-day-old first internodes, being found only in a particulate fraction (Stafford, 1969; Stafford and Baldy, 1970), but I have been unable to demonstrate the activity in incubated excised first internodes, and the activity in green leaves is quite variable (unpublished data). This hydroxylase step appears to be the most ·difficult to demonstrate. In some cases, it has been assayed only in tissue sections. It may be very unstable if released from a membrane localization, a variety of side reactions may interfere with the supply of NADPH, or it may function more effectively in some preparations only if separated from the rest of a multienzyme complex. This activity and its P_{450} status have been more widely studied in etiolated tissues of peas (Russell, 1971) and recently in 3- to 4-day-old etiolated *Sorghum* shoots (Potts, 1973).

The *O*-methyltransferase activity was also detectable in particulate preparations, but it is found mainly in the "soluble" fraction (Stafford and Baldy, 1970). This activity is *Sorghum* might be expected to be similar to that reported for bamboo, which was active with both caffeic and 5-hydroxyferulic acids (Shimada *et al.*, 1972).

4-Hydroxycinnamic Acid Hydroxylase

Recent emphasis in my laboratory has been placed on the step which hydroxylates 4-hydroxycinnamic (*p*-coumaric) to 3,4-dihydroxycinnamic (caffeic) acid (Stafford and Dresler, 1972; Stafford and Bliss, 1973), which was considered the rate-limiting step in internode particulate fractions (Stafford and Baldy, 1970). The natural cofactor has not been determined, but I use ascorbic acid routinely at pH 6, although NADPH and $DMPH_4$ are also effective as electron donors. Leaf preparations contain 10 to 20 times more activity than extracts of first internodes.

In previous work it was postulated that there were at least two hydroxylase activities, one specific for monophenol hydroxylation and the other a classical polyphenolase with both monophenol and diphenol functions. Chlorogenic acid oxidase (CAO) activity (the diphenol function) frequently varies independently of the hydroxylase activity, but no preparation has been obtained that is completely devoid of diphenol oxidase activity. (See Stafford, 1974, for a further discussion of this problem.)

A minimum of three molecular weight forms of 4-hydroxycinnamic acid hydroxylase (HCH) activity in pH 6-buffered extracts have now been separated by ammonium sulfate fractionation and chromatography on Agarose A-15m (Stafford and Bliss, 1973). The HCH activity in seedling tissues and in etiolated leaves was precipitated only above 200 gm/liter

TABLE 3

<small>Changes in the Localization of HCH Activity (Percent) in Ammonium Sulfate
Fractions from Extracts of Etiolated and Green Shoots[a]</small>

	Ammonium sulfate		HCH Total activity (μmoles CA/hr per gm FW)
	0–200 gm/liter	200–500 gm/liter	
First internode—dark grown			
|25°–3 days excised	0	100	0.033
30°—intact-soil		100	0.007
Etiolated coleoptiles and primary leaves	6	94	0.2
Etiolated—leaves[b]	23	77	0.4
Etiolated—leaves plus 6 hours greening[b]	59	41	0.4
Green—2nd and 3rd leaves	97	3[c]	0.4

[a] Stafford and Bliss (1973).

[b] First internodes inhibited by 6 hr white light at 2 days of growth plus 5 days of complete darkness, or plus another 6 hr of white light prior to extraction.

[c] Varies between 3 and 20 percent, activity in this fraction may be unstable.

ammonium sulfate, while the activity of green leaves was precipitated between 0 and 200 gm/liter (Table 3). Six hours of light for greening was sufficient to convert significant amounts to the 0 to 200 gm/liter fraction. Mixing experiments in which etiolated and green leaves were ground together indicated that this difference in response to ammonium sulfate was not an artifact of preparative methods. The amount of activity in green leaves in the 200 to 500 gm/liter fraction was quite variable; it is not clear whether the activity already formed in etiolated leaves is actually attached to a larger particle or whether synthesis is *de novo*.

When the 0 to 200 gm/liter fraction from green leaves was chromatographed on Agarose A-15m, about 90 percent of the activity recovered was found in one peak at the void volume (Fraction I in Fig. 5a), indicating the presence of a very high molecular weight complex. There are still larger complexes or fragments that are trapped on the column. Most of this unrecovered activity, representing approximately 50 percent of the total, can be removed in a microsomal pellet between 37,000 and 100,000g prior to the ammonium sulfate precipitation.

The agarose elution pattern of the 200 to 500 gm/liter ammonium sulfate fraction was very different in etiolated but expanded leaves (Fig. 5b). Ap-

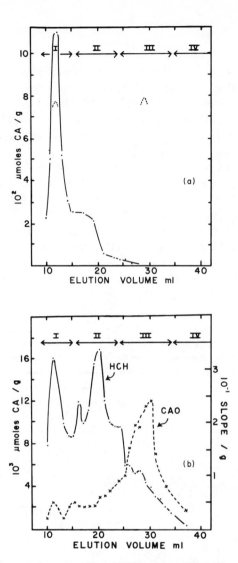

Fig. 5. Agarose A-15 m elution profiles of HCH activity (●—●) and CAO (×---×) activities eluted from 1 × 40 cm columns with 100 mM phosphate, pH 6. The combined fractions I to IV are shown at the top. A sharp peak of peroxidase activity according to visual spot tests was found in I and a broad peak throughout III and IV (Stafford and Bliss, 1973). (a) Activities from green leaves in the 0–500 gm ammonium sulfate fraction, minus the 100,000g pellet. The total HCH activity recovered (158 percent) was equivalent to 0.54 μmole caffeic acid/hr gm fresh weight. Peaks of CAO activity based on visual spot tests are indicated. (b) Activities from etiolated leaves in the 200 to 500 gm ammonium sulfate fraction, minus the 37,000g pellet. The total HCH activity recovered (150 percent) was equivalent to 0.18 μmole/hr gm fresh weight. See Table 4 for percent recovered in each of four peaks. Total CAO activity was equal to a slope of 240/hr gm fresh weight or 222 μmoles O₂/hr gm fresh weight. Approximately 10 percent of the CAO activity was in Fraction I, 25 percent in II, and 64 percent in III.

proximately 25 percent of the recovered activity was found in a peak in the combined Fraction I and about 55 percent in a broader peak in Fraction II. The latter peak may be a complex with three peaks or shoulders, but this has not been studied in detail. Note that the bulk of CAO activity is found in Fraction III, although there is some activity associated with the hydroxylase peaks.

The activity of internode preparations in the 200–500 gm/liter fraction was found in a very small molecular weight fraction (or possibly a retarded peak) called Fraction IV. The profile is not shown in the figures, but the data are summarized in Table 4 together with the shift in the localization of the activity when etiolated leaves are greened. It is necessary to recall that considerable activity was found in the 0–200 gm/liter ammonium sulfate fraction of green tissues.

The above data are based on activities in pH 6 dilute buffer extracts; the results were similar at osmotic concentrations sufficiently high to main-

TABLE 4

Differences in Elution Patterns of HCH Activity from Agarose A-15m Columns of Extracts from Etiolated Shoots and Etiolated and Green Leaves[a]

Tissue	Activity (%) in fractions[b]				Total activity (μmoles CA/hr per gm FW)	Recovered (%)
	I	II	III	IV		
0–200 gm ammonium sulfate/ liter						
Green leaves[c]	90	0	0	0	0.5	150
200–500 gm ammonium sulfate/liter						
First internodes[d]	0	0	0	100	0.02	400
Etiolated coleoptiles + primary leaves	26	43	9	22	0.06	85
Etiolated leaves[e]	25	55	20	0	0.18	150
Etiolated leaves + 6 hr greening[e]	0	90	10	0	0.10	50
Green leaves	0	0	0	0	0	—

[a] Stafford and Bliss (1973).

[b] Fractions in the areas I to IV shown in Figs. 5 and 6 were pooled for analysis or the activities of separate fractions were summed.

[c] 37,000–100,000g pellet analyzed separately; otherwise much of this activity would be trapped on the column.

[d] Internodes from plants grown in soil in the dark at 30°C.

[e] Mainly 2nd and 3rd leaves, primary leaves were discarded. See Fig. 6 for actual elution profile for extract from etiolated leaves.

TABLE 5

<small>Distribution of HCH Activities between Particulate Fractions Obtained Between 500 and 37,000*g* Compared with that of the Supernatant Fraction[a,b]</small>

	pH	O.C.	HCH	
			(% Total activity in 37,000*g* pellet)	Total activity (μmoles caffeic/hr gm fresh wt.)
Green leaves				
MES or PO$_4$	6	L	0–3	0.4
	6	H[c]	3–8	0.22–0.41
PO$_4$	7.4	H[c]	22–82	0.15–0.48
Bicine	8	H[c]	40–80	0.14–0.2[d]

[a] Stafford and Bliss (1973).

[b] Ranges are given for extracts from green leaves and are expressed as % total activity recovered. Tissues were ground in media of low (L) and high (H) osmotic concentrations (O.C.) and various pH values.

[c] Media contained either 0.5 *M* sucrose with or without 3 m*M* MgCl$_2$ with phosphate buffers, or 0.33 *M* mannitol, 2 m*M* KNO$_3$, 2 m*M* Na$_2$EDTA, 2 m*M* ascorbate, 1 m*M* MnCl$_2$, 1 m*M* MgCl$_2$, 0.5 m*M* K$_2$HPO$_4$, 20 m*M* NaCl with 50 m*M* MES (pH 6) or bicine (pH 8) buffers.

[d] Similar results at low osmotic concentration at pH 8.

tain the integrity of organelles such as chloroplasts. At pH 8, however, using Dowex 1 to adsorb phenolics rather than the Polyclar AT used at pH 6, the results were very different for both green and etiolated leaf tissues. Thus, while relatively low amounts of activity were found in a 37,000*g* pellet at pH 6, about 40 to 80 percent of the total activity was associated with such a pellet at pH 8 at either high or low osmotic concentrations (Table 5).

The following speculative interpretation of the above data is my present working hypothesis. HCH activity is part of one or more complexes of enzymes associated with membranes, probably vesicles derived from the endoplasmic reticulum. These HCH activities are more easily solubilized in the internode extracts and a less active protomeric form is isolated. In leaves, the activity remains associated with the vesicles, which in turn may have affinities for various membranes and organelles such as dictyosomes, etioplasts, and chloroplasts. These potential binding sites may be increased at pH 8, although in this case artifacts due to adsorbed phenolics have not been ruled out. Upon grinding at pH 6, a variety of HCH forms of different sizes are released, depending on the membrane localization *in vivo*. Dissociation of polymeric forms may also be involved. Preliminary experiments with Triton-X 100 and butanol treatments lend support to this membrane hypothesis, since experiments with agarose columns indicate that butanol extraction converts Fraction I to IV (H. A. Stafford, unpublished data).

It is tempting to speculate further that these varied molecular weight forms are associated in part with different complexes, leading to the accumulation of caffeic, ferulic, and sinapic acids, respectively. The complex that might be most easily identified is the one responsible for the formation of caffeic acid, since this ester is found in large amounts only in green leaves. At least some of the high molecular weight form found in green leaves might serve this purpose, and a chloroplast localization is suspected, although the complex is not necessarily associated with green lamellae. A chloroplast site would be an efficient one, since a continual supply of NADPH (or ascorbic acid) would be available. In fact, the source of NADPH could be a key factor in the localization of these complexes since this is a cofactor of one and possibly two of the hydroxylation steps. The complexes leading specifically to ferulic and sinapic acids, respectively, will be the hardest to identify, but should be more plentiful in extracts of vascular tissues.

In these cell-free studies, only the non-ester forms have been used as substrates, although esters are often considered to be intermediates (Stafford, 1974). Intermediates may remain bound to the enzyme complex until esterification via UDP-glucose, permitting their transfer to the vacuole, organelles, or other vesicular sites for activation via CoA and ultimate transport to the sites of lignification or flavonoid synthesis.

If multienzyme complex(es) are present *in vivo*, the other enzymes of the sequence must be associated with the aggregate, and kinetic analysis should show that the bound intermediates are better substrates than exogenous precursors (catalytic facilitation, observed within the *arom* complex, Gaertner *et al.*, 1970.) Preliminary experiments indicate that the large molecular weight complex containing HCH activity also contains at least some of the ammonia lyase (PAL and TAL) and cinnamic hydroxylase activities. The latter would be expected since it has been generally found only with microsomal or light membrane fractions (Potts, 1973; Russell, 1971). A recent exception is the activity in dormant potatoes (Camm and Towers, 1973c). The easily solubilized transmethylases must be restudied since they would be found in only some of the complexes. One particular problem could be troublesome. Since the activity of the *in vivo* aggregate may be controlled by the removal of the final product via sequestering in the vacuole, the *in vitro* assay might give only fractional yields unless a means is available to remove the product.

The data presented are only preliminary attempts to associate these activities with one or more multienzyme complexes, and the interpretation is obviously speculative. But they form the basis for a working hypothesis in the hopes of obtaining more sophisticated data, as has been reported for the activities of the *arom* aggregate.

SUPPORTING EVIDENCE FOR THE MULTIENZYME HYPOTHESIS

Although there is no unequivocal direct evidence of the isolation of a multienzyme aggregate or of a specific organelle concerned with the C_6–C_3 hydroxylation sequence in plants, several lines of indirect evidence are available that might be so interpreted.

El-Basyouni and Neish (1966) reported that labeled CO_2, phenylalanine, and tyrosine fed to wheat shoots were incorporated more readily into the phenolic cinnamic acids bound to the ethanol-insoluble residue than into ethanol-soluble fractions. The bound form, cinnamoyl-X, was postulated to be a CoA ester which gave rise to a protein-bound ester. The subsequent hydroxylations took place in this bound form rather than at the level of free acids to produce intermediates in lignin biosynthesis. Zenk and Gross (1972) argue that these results are in conflict with the discovery of cinnamic hydroxylase which acts on cinnamate without carboxyl activation. But another interpretation is that the bound forms of El-Basyouni and Neish were enzyme-bound intermediates of the acids on a multienzyme complex and that catalytic facilitation was involved. If a multienzyme complex is overloaded with exogenous intermediates, free acids might be released, and upon feeding of labeled free cinnamic acids to tissues, incorporation of the label into the ethanol-soluble fraction might occur more readily, as was shown by El-Basyouni and Neish (1966). A major difference in interpretation of these data would be whether activation of the carboxyl group is necessary for the formation of the hydroxylated intermediates. The recent cell-free data for the C_6–C_3 sequence would argue against such a requirement.

There are several examples of correlative increases or coordinate inductions of C_6–C_3 enzymes, an expected phenomenon if complexes are involved. Amrhein and Zenk (1970) report the concomitant induction of phenylalanine ammonia-lyase and cinnamic acid 4-hydroxylase upon illumination of excised hypocotyls of Buckwheat. Similarly, Hahlbrock *et al.* (1971) reported the sequential increase of two groups of related enzymes in cell suspension cultures of parsley, the first involving two of the C_6–C_3 sequence plus the subsequent CoA ligase, followed by the C_{15} sequence. A similar relationship was observed in anthers (Wiermann, 1973). But these activities are not always correlated. Camm and Towers (1973a,b,c) found that only PAL activity was increased in the light in potato slices, although both PAL and cinnamic hydroxylase increased upon aging of the slices. These discrepancies can have other explanations. In first internode extracts of *Sorghum* there is a very active cinnamic hydroxylase prior to incubation of excised internodes, but none is detectable after their subsequent incubation. I interpret these results as evidence of inactivation during isolation, since the incubated internodes are capable of considerable synthesis of phenolic compound requiring the participation of this hydroxylase. A similar disappearance of demonstrable activity has been observed in pea leaves (Russell, 1971).

The presence of multiple forms of an enzyme in the same tissue is also potential evidence of multienzyme complexes in different cellular compartments. An interesting recent example is that of Alibert *et al.* (1972) in *Quercus* root tissues, already discussed by Creasy and Zucker in Chapter 1. They reported that two PAL isozymes could be separated on DEAE-cellulose. Form I, possibly in microbodies, was associated with a degradative pathway leading to C_6–C_1 compounds, while form II, reported to be microsomal, was associated with cinnamic hydroxylase activity. The two lyases could be differentiated on the basis of different sensitivities to inhibition by vanillic acid (I) and by caffeic acid (II). A possible separation of the C_6–C_1 pathways in glyoxysomes, and C_6–C_3 in microsomes and proplastids has also been indicated in *Ricinus* (Kindl and Ruis, 1971).

Two forms of lyases which differ in molecular weight have also been reported by Kindl (1970) in *Hordeum* (barley). He reported the existence of one form with a weight greater than 120,000 with a high activity of TAL relative to PAL, and one smaller than 100,000 with a low activity of TAL relative to PAL. The PAL and TAL activities appear to differ in their sensitivities to C_6–C_3 and C_6–C_1 inhibitors. At least one highly purified form, however, clearly has both PAL and TAL activities (Reid *et al.*, 1972).

The problem of artifacts is a difficult one, especially with multiple molecular weight forms of an enzyme. However, the smaller component should not always be presumed to be the artifact, for aggregation can occur during extraction, and a complex can also be broken down by proteases. The recent isolation studies of phytochrome are a case in point (Rice and Briggs, 1973). Furthermore, dissociation and reassociation of polymeric forms may occur. Schopfer (1971) reported that PAL activity in dilute buffer extracts was associated with a large aggregate (4×10^6 mol wt), while the addition of 0.1 M KCl and 50 gm/liter sucrose converted the aggregate into a smaller form (3×10^5), similar in size to that of highly purified ammonialyases. While he considered the larger form as the artifact due to low ionic strength, one can also interpret his data as the existence of an aggregate that dissociates at the higher salt concentration, a phenomena observed by others (Coleman, 1973; Lin and Key, 1971; Moore, 1973).

While preparations of both microbodies and microsomes have been reported to contain enzyme activities of the C_6–C_3 sequence (Kindl, 1970; Kindl and Ruis, 1971; Ruis and Kindl, 1971; Russell, 1971), there is also evidence of a possible chloroplast localization, but none of these workers eliminated the problem of artifacts produced during grinding. Polyphenol oxidases have been claimed to be associated with chloroplasts for many years. Recently, this localization has been reaffirmed (Czaninski and Catesson, 1972), which is of considerable interest in the light of claims of the possible presence and synthesis of flavonoids and C_6–C_3 complexes (Kannagara *et al.*, 1971; Oettmeir and Heupel, 1972; Weissenböck *et al.*, 1972). Recently PAL and TAL activities have been reported in highly purified chloroplast and etioplast preparations from *Hordeum* (unpublished work discussed by McClure, 1974). A 4-hydroxycinnamic acid hydroxylase activity has been reported in chloroplast preparations by Bartlett *et al.* (1972) and Parish (1972).

None of the studies conducted so far unequivocably pinpoints the intracellular localization of any of the enzymes of the C_6–C_3 sequence, and their association with a variety of particles and the cytosol could be artifacts of random adsorption. However, adsorption or binding may have physiological significance (Kuczenski, 1973), and multiple sites of localization might be expected if there are a series of complexes as postulated in Fig. 4.

Potential Multienzyme Complexes and Lignin Formation

DEFINITION AND CRITERIA FOR IDENTIFICATION AND ASSAYS

The precise molecular and three-dimensional structure of lignin is still not known, although it is generally agreed that it is a polymer of phenylpropane units found in the cell walls of higher plants (Brown, 1966). Isotopic tracer work has clearly indicated that the C_6–C_3 compounds discussed above are actively incorporated into this wall-localized polymer (El-Basyouni and Neish, 1966; Higuchi, 1971; Shimada, 1972). But the definition of lignin is too rigidly based on woody plants, and the only quantitative assays for lignin that measure the total amount of the polymer (such as the weight method used in Klason lignin analyses) are not appropriate for herbaceous materials. Other assays, such as the phloroglucinal or Maüle reagents, are based on detection of only one functional group and do not distinguish between monomeric and polymeric forms (Sarkanen and Ludwig, 1971; Stafford, 1962).

Lignin in grasses may be considered to consist of one to three types: (1) A classical or core lignin similar to that of woody plants made up of C_6–C_3 units in which the terminal carbon of the side chain is reduced to the aldehyde and the alcohol prior to oxidative polymerization; in grasses these units are made up of all three hydroxylation states, i.e., the monophenol, diphenol, and triphenol, the hydroxyl at the 3- and 5-positions being methylated. (2) Grasses are characterized by a high proportion of the acid derivatives such as 4-hydroxycinnamic and ferulic acids in ester linkages to carbohydrate or to the lignin polymer itself (Higuchi, 1971; Shimada et al., 1971). (3) Since peroxidase can polymerize ferulic acid directly into a ligninlike product, and saponification and esterase activities cannot remove the bulk of this material, I believe these acids can exist in linkages in addition to that of simple ester groups (Stafford, 1962). The terminal acid groups need not exist as free acids in the final wall polymer, but may be subsequently modified and linked to other molecules.

REDUCTIVE PATHWAY TO LIGNIN PRECURSORS

When the various hydroxycinnamic acid derivatives are reduced to aldehydes or alcohols, the terminal carboxyl group must be activated (Zenk and Gross, 1972). Recently, CoA ligases (Hahlbrock and Grisebach, 1970) and the subsequent reduction of p-coumaryl:CoA and feruloyl:CoA to alcohols via NADPH has been demonstrated by two groups (Ebel and Grisebach, 1973; Gross et al., 1973). This pathway is discussed by Grisebach in this volume.

Once the aldehyde and alcohol phenylpropane precursors are formed, presumably within some vesicular structure to ensure isolation from degradative enzymes, these precursors, and possibly the free acids also, must come in contact at the appropriate time and site with peroxidase and H_2O_2. While it is now overwhelmingly clear that peroxidase is the polymerizing catalyst (Harkin and Obst, 1973), the identification of the particular isozymes involved, as well as the source of the H_2O_2, or the site of the oxidative polymerization are not known. It is also not clear what controls the formation of linkages of the lignin core to the hemicellulose matrix and to the hydroxycinnamic acids.

THE INTRACELLULAR SITE OF PEROXIDASE

A large number of reports indicate both a wall and particulate localization as well as a "soluble" component for peroxidases in cell-free extracts (Stafford and Bravinder-Bree, 1972). Electron microscope data also indicate both a wall as well as a widespread membrane localization of peroxidases in organelles, plasmalemma, and tonoplast (Hall and Sexton, 1972; Harkin and Obst, 1973; Hepler *et al.*, 1972; Nougarède, 1971; Parish, 1972). Artifacts are a serious problem in microscopy of fixed tissues. The stain diaminobenzidine (DAB) is photooxidized, making any chloroplast localization difficult to interpret (Hepler *et al.*, 1972). Furthermore, catalase can act as a peroxidase, and even polyphenoloxidase is a weak pseudoperoxidase (Jolley *et al.*, 1973). Some of the above investigators have used inhibitors to differentiate between these possibilities.

Detailed accounts of the sequence of cytoplasmic events during lignification in xylem have been given by Robards and Kidwai (1969), Roberts (1969), and Hepler *et al.* (1970). Dictyosomes (golgi) and their associated vesicles which stain positively for peroxidase with DAB are postulated to be vehicles of transport to the site of lignification. Cytoplasmic vesicles fuse with the plasmalemma in the area of active secondary wall thickening. Isotopic labels from lignin precursors fed to tissues are found associated with dictyosomes and with vesicles adjacent to microtubules during active secondary wall formation (Pickett-Heaps, 1968). Earlier studies also indicated the presence of cytoplasmic constituents with phloroglucinol-staining properties when tissues were incubated with phenolic substrates and H_2O_2 (Stafford, 1960).

Multiple forms of peroxidase are commonly reported. A typical array of such multiple bands of activities (stained with *o*-dianisidine) can be observed after electrophoresis of *Sorghum* extracts on Sepraphore III at pH 8.3 (Stafford and Bravinder-Bree, 1972), (Fig. 6). Although none of the forms was restricted to a wall localization, the cathodic group was associated with walls, using cation exchange and infiltration-centrifugation techniques that eliminate artifacts. However, the dominant form, C_3, was found mainly in the cortex, and was not detectable in the walls of the stele, the only tissue to give a classical phloroglucinol test for lignin. The C_3 form, therefore, presumably catalyzes one of the many other chemical reactions typical of peroxidases acting either peroxidatically or oxidatively. Since

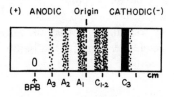

Fig. 6. Electrophoretic pattern of peroxidase isozymes in incubated internodes. Sepraphore III strips were stained with *o*-dianisidine after electrophoresis for 2 hours in borate buffer at pH 8.3. BPB, bromophenol blue (Stafford and Bravinder-Bree, 1972).

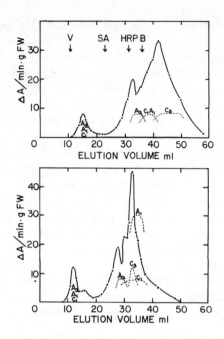

Fɪɢ. 7. Agarose A-15 m elution profiles of peroxidase activity in extracted whole inter-nodes (upper graph) and of steles from incubated internodes (lower graph). Dotted line peaks are visual estimates of isozyme constituents. V, Initial void volume; SA, serum albumin; HRP, horseradish peroxidase; B, Bromcresol green (Stafford and Bravinder-Bree, 1972).

one or more of these other forms is presumably involved in lignification, the suspected isozyme would be in the C_{1-2} area, since it is found in both the cytoplasm and the wall of the stele. Some of these forms existed in a very high molecular weight complex (Fraction I from Agarose A-15m columns) (Fig. 7). It should be noted that bands with similar mobilities also appeared in the lower molecular weight fraction, and that there were marked differences in the patterns found between the cortex (reflected in the profile for the whole internodes) and the stele.

Finally, it is necessary to emphasize that tissues such as the cortex that normally do not produce classical lignins can be forced to do so by feeding exogenous precursors (Stafford, 1967).

H_2O_2-Pʀᴏᴅᴜᴄɪɴɢ Sʏsᴛᴇᴍs

Since ligninlike compounds are produced peroxidatically, the natural source of H_2O_2 is of vital importance, but little is known concerning this.

The classical system to suspect would be a flavoprotein, but no specific one has been identified. Two other possibilities should be considered. One is that hydroxylase enzymes, in the absence of phenolic substrates but in the presence of reductants such as NADPH, yield H_2O_2 in an uncoupled system (Boveris *et al.*, 1972). Another possibility is a polyamine oxidase found in leaves and in seedlings of barley and maize which is reported to be localized in the vascular tissue (Smith, 1971). Boveris *et al.* (1972) have determined physiological levels of H_2O_2 production in liver cell fractions. Microsomes had the highest physiological level of H_2O_2, with peroxisomes a close second. Such techniques need to be used with plant tissues.

OXIDATIVE POLYMERIZATION TO LIGNIN IN CELL-FREE SYSTEMS

The formation of the DHP (dehydrierungspolymerizat) of Freudenberg is of limited value since long incubation times are required, but some use of this technique has been made (Sarkanen and Ludwig, 1971; Shimada *et al.*, 1971). A far more productive system is the inert macromolecular matrix system first introduced by Siegel (1957). Using his paper-peroxidase system, Stafford (1964) found that ferulic acid was a good precursor of ligninlike products as well as coniferyl alcohol. This technique of Siegel's should be reexamined with modern chromatographic matrices and bound peroxidase. A third approach using cell suspension cultures has been marred by difficulties in assessing criteria for lignin detection. The identification of lignin as a major component of the macromolecular fibril secreted from cell suspension cultures has been withdrawn (Leppard *et al.*, 1971; Leppard and Colvin, 1971). Another recent report needs substantiation (Moore, 1973). This is a potentially exciting approach because of the secretion of macromolecules containing peroxidase activity (Olson *et al.*, 1969).

Conclusions: A Speculative Overview of the Role of Multienzyme Complexes in the Metabolism of Phenolic Compounds

My present working hypothesis, based on fragmentary enzymological evidence of my own and that published by others, and on the interpretation of subcellular events during xylary differentiation reported by electron microscopists, is summarized in Fig. 8. Any consideration of the conversion of aromatic amino acids to the phenylpropane units that give rise to lignin and flavonoids must take into account the dynamic interactions between cellular constituents. Organelles are in constant intermittent contact with each other and with the cytoplasmic membranes, including the endoplasmic reticulum. The phase-contrast films of Wildman's research

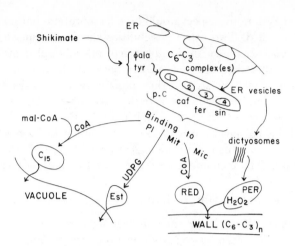

Fig. 8. Overall view of subcellular and biochemical events in the formation of C_6–C_3 phenolic compounds and the accumulation of esters (Est) and flavonoids (C_{15}) in vacuoles, and lignins (C_6–C_3)$_n$ in the cell wall. ER, endoplasmic reticulum; mal-CoA, Malonyl-CoA; Pl, plastid; Mit, mitochondria; Mic, microbodies; RED, carboxyl-reducing enzymes; PER, peroxidase; other abbreviations as for Fig. 1.

group (Department of Botanical Sciences, U.C.L.A., available through the Extension Services at the University of California at Los Angeles) indicate dramatically the continual changes in shape, fragmentations, and even apparent fusions of mitochondria and chloroplasts. In particular, the outer portion of the chloroplast is maintained in a highly mobile state.

One or more types of multienzyme complexes of the C_6–C_3 sequence might be budded off as vesicles from the endoplasmic reticulum (ER) and become attached to various organelles. A steady source of NADPH is required for several key steps in the basic C_6–C_3 sequence and in the subsequent changes to produce flavonoids and lignins. This could come from plastids and mitochondria as well as the "soluble" isocitrate dehydrogenase and pentose shunt system. Since a CoA source is also needed for carboxyl activation and for flavonoid biosynthesis, a key binding target could be the outer membranes of chloroplasts and mitochondria, where fusion of the vesicles with others containing the C_{15}-synthesizing enzymes might occur. Upon glucosylation, the particles containing flavonoids and C_6–C_3 esters might be secreted into the vacuole. Other C_6–C_3 vesicles might attach to the dictyosomes, coming in contact with particles assembled there containing the reductive sequence and the peroxidase-H_2O_2 system. These vesicles are believed to be channeled to the wall by microtubles. Then the single or possibly fused vesicular bodies containing both peroxidase and lignin substrates would be transported across the plasmalemma into the cell wall area, where attachment to existing carbohydrate would occur, with subsequent or continued oxidative polymerization.

This is my present working hypothesis to explain the dynamic changes occurring within the cytoplasm and the adjoining wall. But phase and electron microscopy combined with sophisticated enzymological techniques involving multienzyme systems and membrane-oriented particles will be necessary to draw a more accurate picture.

REFERENCES

Alibert, G., R. Ranjeva, and A. Boudet. 1972. *Biochim. Biophys. Acta* **279**:282–289.
Amrhein, W., and M. H. Zenk. 1970. *Naturwissenschaften* **57**:312–313.
Avery, G. S. 1930. *Bot. Gaz. (Chicago)* **89**:1–39.
Bartlett, D. J., J. E. Poulton, and V. S. Butt. 1972. *FEBS Lett.* **23**:265–267.
Boveris, A., N. Oshino, and B. Chance. 1972. *Biochem. J.* **128**:617–630.
Brown, S. A. 1966. *Annu. Rev. Plant Physiol.* **17**:223–244.
Camm, E. L., and G. H. N. Towers. 1973a. *Can. J. Bot.* **51**:824–825.
Camm, E. L., and G. H. N. Towers. 1973b. *Phytochemistry* **12**:961–973.
Camm, E. L., and G. H. N. Towers. 1973c. *Phytochemistry* **12**:1575–1580.
Coleman, R. 1973. *Biochim. Biophys. Acta* **300**:1–30.
Craker, L. E., and P. J. Wetherbee. 1973. *Plant Physiol.* **51**:436–438.
Czaninski, Y., and A. Catesson. 1972. *J. Microsc. (Paris)* **15**:409.
Doggett, H. 1970. "Sorghum." Longmans, Green, New York.
Downs, R. J., and H. W. Siegelman. 1963. *Plant Physiol.* **38**:25–30.
Ebel, J., and H. Grisebach. 1973. *FEBS Lett.* **30**:141–143.
El-Basyouni, S. Z., and A. C. Neish. 1966. *Phytochemistry* **5**:683–691.
Engelsma, G. 1972. *Plant Physiol.* **50**:599–602.
Gaertner, F. H., M. C. Erickson, and J. A. DeMoss. 1970. *J. Biol. Chem.* **245**:595–600.
Ginsburg, A., and E. R. Stadtman. 1970. *Annu. Rev. Biochem.* **39**:429–472.
Gross, G. G., J. Stockigt, R. L. Mansell, and M. H. Zenk. 1973. *FEBS Lett.* **31**:283.
Hahlbrock, K., and H. Grisebach. 1970. *FEBS Lett.* **11**:62–64.
Hahlbrock, K., B. Abel, R. Ortmann, A. Sutter, E. Willmann, and A. Grisebach. 1971. *Biochim. Biophys. Acta* **244**:7–15.
Hall, J. L., and R. Sexton. 1972. *Planta* **108**:103–120.
Harborne, J. B. 1973. *In* "Phytochemistry," Vol. II, pp. 344–380. Van Nostrand Reinhold, New York.
Harkin, J. M., and J. R. Obst. 1973. *Science* **180**:296–297.
Hepler, P. K., A. E. Fosket, and E. H. Newcomb. 1970. *Amer. J. Bot.* **57**:85–96.
Hepler, P. K., R. M. Rice, and W. A. Terranova. 1972. *Can. J. Bot.* **50**:977–983.
Higuchi, T. 1971. *Advan. Enzymol.* **34**:207–283.
Hinchman, R. R. 1972. *Amer. J. Bot.* **59**:805–818.
Jacobson, J. W., B. A. Hart, C. H. Day, and N. H. Giles. 1972. *Biochim. Biophys. Acta* **289**:1–12.
Jolley, R. L., L. H. Evans, and H. S. Mason. 1973. *J. Biol. Chem.* **249**:335.

Kannangara, C. G., K. W. Henningsen, P. K. Stumpf, and D. von Wettstein. 1971. *Eur. J. Biochem.* **21**:334–338.

Kindl, H. 1970. *Hoppe-Seyler's Z. Physiol. Chem.* **351**:792–798.

Kindl, H., and H. Ruis. 1971. *Phytochemistry* **10**:2633–2636.

Kuczenski, R. T. 1973. *J. Biol. Chem.* **248**:5074–5080.

Leppard, G. C., and J. R. Colvin. 1971. *J. Polym. Sci., Part C* **36**:321–326.

Leppard, G. C., J. R. Colvin, D. Rose, and S. M. Martin. 1971. *J. Cell Biol.* **50**:63.

Lin, C. Y., and J. L. Key. 1971. *Plant Physiol.* **48**:547–552.

McClure, J. W. 1974. *In* "The Flavonoids" (J. B. Harborne and T. J. Mabry, eds.). Chapman & Hall, London (in press).

Mohr, H. 1972. "Lectures on Morphogenesis." Springer-Verlag, Berlin and New York.

Moore, T. S., Jr. 1973. *Plant Physiol.* **51**:529–536.

Nougarède, A. 1971. *C.R. Acad. Sci.* **273**:864–867.

Oettmeier, W., and A. Heupel. 1972. *Z. Naturforsch. B* **27**:177–182.

Olson, A. C., J. J. Evans, A. P. Frederich, and E. F. Jansen. 1969. *Plant Physiol.* **44**:1594–1600.

Parish, R. W. 1972. *Eur. J. Biochem.* **31**:446–455.

Pickett-Heaps, J. A. 1968. *Protoplasma* **65**:181–205.

Potts, J. R. M. 1973. Ph.D. Thesis, University of California, Davis. [*J. Biol. Chem.*, 1974 (in press).]

Reid, P. D., E. A. Havir, and H. V. Marsh, Jr. 1972. *Plant Physiol.* **50**:480–484.

Rice, H. V., and W. R. Briggs. 1973. *Plant Physiol.* **51**:927–938.

Robards, A. W., and P. Kidwai. 1969. *New Phytol.* **68**:343–349.

Roberts, L. W. 1969. *Bot. Rev.* **35**:201–250.

Ruis, H., and H. Kindl. 1971. *Phytochemistry* **10**:2627–2631.

Russell, D. W. 1971. *J. Biol. Chem.* **246**:3870.

Sarkanen, K. V., and C. H. Ludwig. 1971. "Lignins." Wiley (Interscience), New York.

Schopfer, P. 1971. *Planta* **99**:339.

Schopfer, P., and H. Mohr. 1972. *Plant Physiol.* **49**:8–10.

Sedgwick, B. 1973. *Phytochem. Soc. Symp.* **9**:178–217.

Shimada, M. 1972. *Wood Res.* **53**:19.

Shimada, M., T. Fukuzuka, and T. Higuchi. 1971. *Tappi* **54**:72.

Shimada, M., H. Fushiki, and T. Higuchi. 1972. *Phytochemistry* **11**:2657.

Siegel, S. M. 1957. *J. Amer. Chem. Soc.* **79**:1628.

Smith, T. A. 1971. *Biochem. Biophys. Res. Commun.* **41**:1452.

Stafford, H. A. 1960. *Plant Physiol.* **35**:612.

Stafford, H. A. 1962. *Plant Physiol.* **37**:643.

Stafford, H. A. 1964. *Plant Physiol.* **39**:350.

Stafford, H. A. 1965. *Plant Physiol.* **40**:130.

Stafford, H. A. 1966. *Plant Physiol.* **41**:953.

Stafford, H. A. 1967. *Plant Physiol.* **42**:450–455.

Stafford, H. A. 1969. *Phytochemistry* **8**:743.

Stafford, H. A. 1970. *Phytochemistry* **9**:1799.

Stafford, H. A. 1974. *Annu. Rev. Plant Physiol.* **25**:459–486.

Stafford, H. A., and R. Baldy. 1970. *Plant Physiol.* **45**:215.

Stafford, H. A., and M. Bliss. 1973. *Plant Physiol.* **52**:453.

Stafford, H. A., and S. Bravinder-Bree. 1972. *Plant Physiol.* **49**:950.

Stafford, H. A., and S. Dresler. 1972. *Plant Physiol.* **49**:590–595.

Sumper, M., and F. Lynen. 1972. *In* "Protein–Protein Interactions" (R. Jaenicke and E. Helmreich, eds.), pp. 364–394. Springer-Verlag, Berlin and New York.

Weissenböck, G., I. Fleing, and H. G. Ruppel. 1972. *Z. Naturforsch. B* **27**:1216.

Wiermann, R. 1973. *Planta* **110**:353.

Zenk, M. H., and G. G. Gross. 1972. *Recent Advan. Phytochem.* **4**:87.

Zucker, M. 1972. *Annu. Rev. Plant Physiol.* **23**:133.

PHOTOREGULATION OF PHENYLPROPANOID AND STYRYLPYRONE BIOSYNTHESIS IN *Polyporus hispidus*

G. H. N. TOWERS, C. P. VANCE, and A. M. D. NAMBUDIRI

Botany Department, University of British Columbia,
Vancouver, British Columbia

Introduction

The phenolic substances of microorganisms, including fungi, are derived mainly from (a) acetate and malonate via the polyketide pathway or from (b) intermediates in the shikimate pathway including phenylpyruvic and hydroxyphenylpyruvic acids. Among the fungi, however, the Basidiomycetes and some Ascomycetes are able to deaminate, nonoxidatively, phenylalanine and tyrosine to cinnamate and *p*-coumarate, a metabolic feature which they share with plants. This avails them of another type of molecular unit for the elaboration of complex phenols.

In higher plants, hydroxy- and methoxycinnamic acids as well as the corresponding aldehydes and alcohols are the substrates for a vast network of synthetic reactions. So far, it would appear that fungi are less talented in this respect, there being less than 20 compounds known from Basidiomycetes which can be derived from cinnamic acid (see Towers, 1969). It is probably true to say, however, that there has been very little systematic screening of higher fungi for these compounds.

With our discovery that certain fungi contain phenylalanine ammonialyase (PAL) and tyrosine ammonia-lyase (TAL) activity (Bandoni *et al.*, 1968; Power *et al.*, 1965), it was decided to make a detailed study of phenylpropanoid metabolism in these organisms. Fungi are generally easy to grow, they often yield active enzyme preparations quite readily, and finally, from a comparative standpoint, it is of interest to discover to what extent their systems of aromatic metabolism resemble those of green land plants. Around 1965, with P. V. Subba Rao and Keith Moore, a start was made on the problem. Our work with yeastlike Basidiomycetes such as *Sporobolomyces roseus* (Moore *et al.*, 1967, 1968), *Schizophyllum commune* (Moore and Towers, 1967), and *Ustilago hordei* (Subba Rao *et al.*, 1967) showed that phenylalanine could be degraded to CO_2 by these fungi through a "cinnamate pathway," the sequence being

Phenylalanine $\rightarrow$ cinnamate $\rightarrow$ benzoate $\rightarrow$ *p*-hydroxybenzoate
$\rightarrow$ protocatechuate $\rightarrow CO_2$

This is shown in Fig. 1. Evidence for this scheme was based on the formation

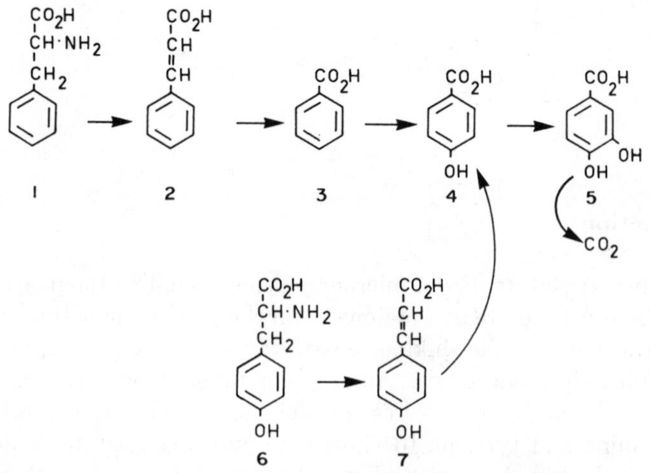

FIG. 1. The "cinnamate" pathway for the catabolism of phenylalanine and tyrosine in certain Basidiomycetes.

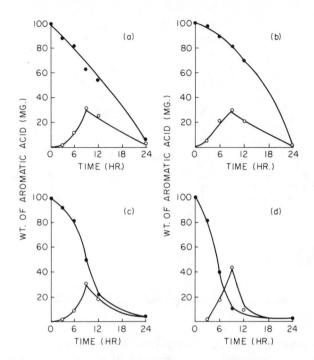

Fig. 2. Time course of the disappearance of substrate (●) and formation of proto-catechuic acid (○) in the presence of *Sporobolomyces roseus*. The substrates used were (a) cinnamic acid; (b) benzoic acid; (c) *p*-hydroxybenzoic acid; (d) *p*-coumaric acid. Washed cells (3 gm wet weight) were added aseptically to a flask containing 100 mg of substrate in 0.01 M Na phosphate buffer pH 7 (100 ml) and incubated at 25°C.

of subsequent members of the pathway when an earlier compound in the sequence was provided to the intact organism (Fig. 2). Further evidence for this metabolic scheme was obtained in tracer work. For example, $^{14}CO_2$ was produced by *Schizophyllum commune* when growing cultures were provided with ring-labeled phenylalanine, cinnamic acid, or benzoic acid (Moore and Towers, 1967). Ring-cleavage of protocatechuate would be expected to occur via a 3,4-oxygenase. This enzyme has been identified by Cain *et al.* (1968) in *Sporobolomyces* and other fungi. We were able to partially purify an enzyme from cultures of the Basidiomycete, *Tilletiopsis washingtonensis*, which catalyzes endiol ring-cleavage not only of protocatechuate but of homoprotocatechuate and caffeate as well (Subba Rao *et al.*, 1971).

Tyrosine appeared to be degraded through *p*-coumarate in a similar sequence in some of these fungi. We did not establish, however, whether this was the main catabolic route for the degradation of tyrosine or

Fɪɢ. 3. Other possible degradative routes for phenylalanine and tyrosine.

phenylalanine in these Basidiomycetes. Figure 3 indicates other possibilities. In all fungi which we examined, the hydroxyphenylacetic acids were products of phenylalanine and tyrosine metabolism but whether these acids are ultimately degraded to CO_2 remains to be determined. This subject is discussed in a review by Towers and Subba Rao (1972).

Phenolic Metabolism in *Polyporus*

Hɪsᴘɪᴅɪɴ Bɪosʏɴᴛʜᴇsɪs

We turned our attention to a Basidiomycete, namely *Polyporus hispidus*, that appeared to metabolize cinnamic acids more in the manner of a higher plant. *P. hispidus* is a bracket fungus parasitic on deciduous trees including *Fraxinus* and *Quercus*. The fruiting bodies of *P. hispidus* and *P. schweinitzii* as well as species of *Gymnopilus* had been shown to contain styrylpyrones very similar indeed to the styrylpyrones of higher plants (Bu'Lock and Smith, 1961; Edwards *et al.*, 1961; Ueno *et al.*, 1964; Hatfield and Brady, 1968, 1969, 1971; Brady and Benedict, 1972). Moreover, one of these compounds, hispidin, 6-(3,4-dihydroxystyryl)-4-hydroxy-2-pyrone, (16, Fig. 4) was produced when the fungus was grown in sterile culture (Bu'Lock, 1967). These compounds appear to be derived by an extension of the side chain of a cinnamyl derivative with two acetate equivalents.

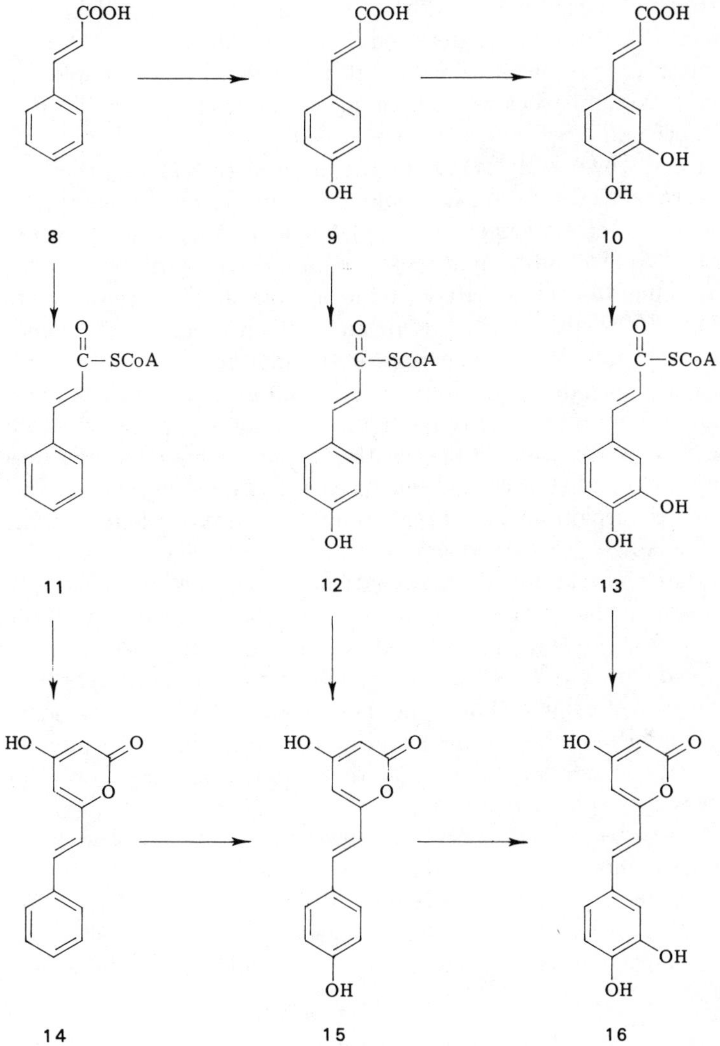

FIG. 4. Alternate routes proposed for the biosynthesis of hispidin in *Polyporus hispidus*: **8**, cinnamic acid; **9**, *p*-coumaric acid; **10**, caffeic acid; **11**, cinnamoyl-coenzyme A; **12**, *p*-coumaroyl-coenzyme A; **13**, caffeoyl-coenzyme A; **14**, styrylpyrone; **15**, bis-noryangonin; **16**, hispidin. Perrin (1972).

A study of hispidin biosynthesis was initiated in our laboratory by Perrin (1972). Hispidin was detected in the medium of *P. hispidus* about 10 days after growth in stationary culture with glucose as main carbon source. Maximum yield was 0.6 μg/mg tissue. Before hispidin appeared, *p*-coumaric and later caffeic acid could be detected in the culture medium (Perrin and Towers, 1973a). Administration of tracers showed that these acids were labeled from phenylalanine, and in older cultures, there was a relatively greater conversion of phenylalanine to caffeic acid. Tracer studies in relation to hispidin biosynthesis were also carried out and it was shown that *p*-coumarate and caffeate were the best precursors of the aromatic ring (Perrin and Towers, 1973b). Acetate and malonate, on the other hand, were incorporated into the pyrone ring of the molecule. In other words, the formation of hispidin from cinnamic acid and malonate, probably through the CoA esters, was established (Fig. 4). We were pleased to learn that similar results had been obtained with *Polyporus schweinitzii*. Hatfield and Brady (1973) had shown good incorporation of labeled phenylalanine and acetate into hispidin in this species and informed us of their results.

At this stage, we turned our attention to another aspect of hispidin biosynthesis. Preliminary investigations had shown that the yellow pigmentation characteristic of hispidin production did not develop in cultures of *P. hispidus* maintained in continuous darkness. When these 2-week old dark-grown cultures were exposed to fluorescent light, however, they developed color within 24 hours. Light obviously was, in some way, controlling hispidin biosynthesis and we examined this effect more closely, particularly in relation to some of the enzymes concerned with phenyl-alanine and tyrosine metabolism.

The enzymes which were examined all relate to phenylpropanoid metabolism and some to hispidin biosynthesis. They are (a) the ammonia lyases for phenylalanine and tyrosine; (b) the aminotransferases for these amino acids; (c) cinnamate- 4-hydroxylase (d) *p*-coumarate hydroxylase and (e) *p*-coumarate:CoA ligase. Brief descriptions of these are given below.

Phenylalanine and Tyrosine Ammonia-Lyases

Relatively crude preparations, obtained by ammonium sulfate precipita-tion and subsequent Sephadex G-25 filtration of buffer extracts, were used. Activities obtained from this organism were somewhat low when compared with preparations we had obtained from some other fungi, e.g., *Sporobolo-myces roseus* (Camm and Towers, 1969). Whether or not the activities for phenylalanine and tyrosine reside in one enzyme could not be determined, but the relative enzyme activities for the two amino acids were found to differ depending on the conditions under which the fungus was grown.

Phenylalanine and Tyrosine Aminotransferases

These were the same preparations as those used for the ammonia lyase determinations. A modification of the assay described by Diamondstone (1966), based on the alkali-catalyzed oxidation of *p*-hydroxyphenyl-pyruvate to *p*-hydroxybenzaldehyde and oxalate, was used.

Cinnamate- 4-Hydroxylase

Over 95 percent of the activity of this enzyme was located in the micro-somal fraction, with remaining activity being present in mitochondria (Vance *et al.*, 1973). The enzyme, which required NADPH and FAD for maximum activity, did not catalyze the hydroxylation of benzoic acid; it differs from the enzyme of higher plants in its apparent requirement for FAD.

p-Coumarate Hydroxylase

In the early stages of purification, the preparations were found to contain caffeic acid oxidase activity in addition to *p*-coumaric acid hydroxylase activity, indicating that it is a tyrosinase type (EC 1.10.3.1) of enzyme. Gel filtration using Sephadex G-200, however, resulted in a resolution into two active fractions designated as E_1 and E_2, the latter exhibiting no detectable activity towards caffeic acid. Unlike the enzyme described for *Streptomyces* (Nambudiri *et al.*, 1972), which also hydroxylates tyrosine, and phloretic (4-hydroxydihydrocinnamic) and *p*-hydroxyphenylacetic acids, the enzymes from *P. hispidus* displayed a narrow range of specificity.

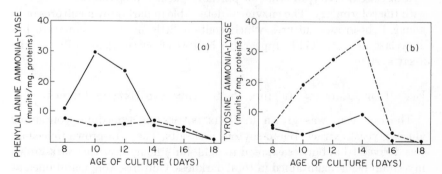

FIG. 5(a). Effect of light on phenylalanine ammonia-lyase activity in *Polyporus hispidus*. Phenylalanine ammonia-lyase activity was assayed in the cell-free extracts derived from dark-grown (– – –) and light-grown (——) mycelium at different days of growth.

FIG. 5(b). Effect of light on tyrosine ammonia-lyase activity in *Polyporus hispidus*. Tyrosine ammonia lyase activity was assayed in the cell-free extracts derived from dark-grown (– – –) and light-grown (——) mycelium at different days of growth.

Both E_1 and E_2 were also effective in the hydroxylation of bis-noryangonin to give hispidin. Precise measurements of velocity of the reaction with bis-noryangonin as substrate were not possible since both substrate and product appear to bind to the protein to some extent, rendering spectrophotometric determination difficult. All evidence, however, indicated that the activities for p-coumarate and bis-noryangonin reside in the same protein. Unlike the phenolase from spinach (Vaughn *et al.*, 1969), the enzyme failed to hydroxylate flavonoids such as kaempferol and apigenin, compounds which have a p-hydroxyphenyl group.

Although ascorbic acid served as a reductant, NADPH and NADH proved to be more effective. Of a number of compounds tested, ferulic acid was the only one causing a significant inhibition (30 percent). The elution patterns of E_1 and E_2 on a calibrated Sephadex G-200 column indicated that the two enzymes had approximate molecular weights of 185,000 and 45,000, respectively. This suggests that E_1 is a tetramer of E_2. Phenolases (tyrosinases) are known to exist in multiple forms ranging from monomers to higher oligomers (Malstrom and Ryden, 1968). For example, the tyrosinase from *Aspergillus nidulans* exists in monomer–tetramer equilibrium (Bull and Carter, 1973).

p-Coumarate:CoA ligase

Difficulty was encountered in detecting this enzyme using the procedure of Walton and Butt (1970). This procedure involves the conversion of the product (the CoA ester) to the hydroxamate which is determined with ferric chloride. The formation of caffeic acid, as a result of p-coumarate hydroxylase activity, in crude or partially purified preparations, interferes with the colorimetry. The enzyme is detectable in dark-grown cultures or in young light-grown cultures which contain little or no p-coumarate hydroxylase activity. GTP appears to be less effective than ATP in the assay system.

EFFECT OF LIGHT ON HISPIDIN PRODUCTION AND ENZYME LEVELS

The organism was grown in Roux bottles on a medium containing glucose, yeast extract, Bacto-soytone and salts. The light-grown mycelium was derived from bottles exposed to light for $\frac{1}{2}$ hour/day, the dark-grown mycelium being maintained in total darkness. Cultures were maintained at 25°C in incubators provided with daylight fluorescent tubes (Nambudiri *et al.*, 1973).

There was no formation of pigment in the dark-grown mycelium even after 18 days of growth. When exposed to light, pigment was detected in the mycelium after 10 days of growth, reaching a maximum value in 16 days

(0.57 μg/mg of tissue). PAL activity was considerably higher in light-grown cultures and TAL activity was maximal in dark-grown cultures (Fig. 5). The activities of transaminases for phenylalanine and tyrosine were higher in dark-grown cultures. Cinnamic acid hydroxylase activity, estimated by determining the amount of radioactive *p*-coumarate formed after administration of phenylalanine-^{14}C to cultures, was found to reach a maximum at 12 days and was higher in the light-grown mycelium than in the dark-grown culture. Elaboration of *p*-coumarate hydroxylase and and the appearance of hispidin followed a similar pattern (Fig. 6 and 7). Neither pigment nor hydroxylase activity was detected in mycelium grown in the dark. Finally, *p*-coumarate:CoA ligase activity could be detected both in light- and dark-grown cultures.

These results are summarized in the diagram shown in Fig. 8. There is a clear-cut effect of light on the cinnamate pathway leading from phenylalanine to hispidin. Because *p*-coumarate hydroxylase activity paralleled the synthesis of hispidin these two activities were next examined in greater detail.

The first questions we asked were (a) How much light is required to trigger these activities? (b) What kind of light is necessary?

We were fortunate in interesting Dr. E. B. Tregunna in the project and he assisted in the design and implementation of the following experiments.

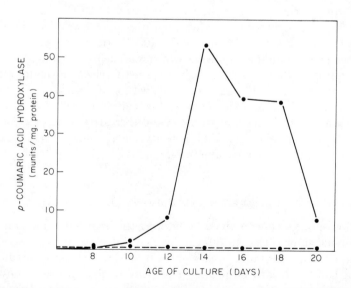

Fig. 6. Effect of light on *p*-coumaric acid hydroxylase activity in *Polyporus hispidus*; (- - -) dark-grown and (——) light-grown mycelium.

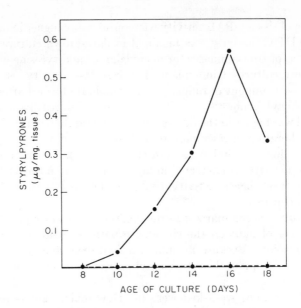

Fɪɢ. 7. Effect of light on pigmentation in *Polyporus hispidus* (– – –) dark-grown and (——) light-grown mycelium.

Cultures were grown in light-proof incubators at 25°C for 13 days, after which they were exposed to varying intensities and/or wavelengths of light. The light source was a Philips 12 V, 100 W tungsten iodide filament, mounted in a 35 mm slide projector which was enclosed in a lightproof box.

The effective dosage of white light required to elicit maximum response was found to be 5×10^4 ergs/cm² provided for 10 seconds. *p*-Coumarate hydroxylase activity and pigment development appeared within 12 hours, reaching a maximum level in 16 hours and thereafter remaining constant; the patterns were identical. PAL activity, on the other hand, reached a maximum in 12 hours. Cinnamate 4-hydroxylase doubled in the 24 hour period.

Action Spectra for Enzyme and Pigment Development

Action spectra were obtained by filtering the light sources with a wedge interference filter (Zeiss, 4663074). The light intensity at each wavelength used in the study was measured with an Isco Model 3R Spectroradiometer with a half-bandwidth of 15 nm. The stray light response to unwanted wavelengths was 0.01 percent. The action spectra for styrylpyrone

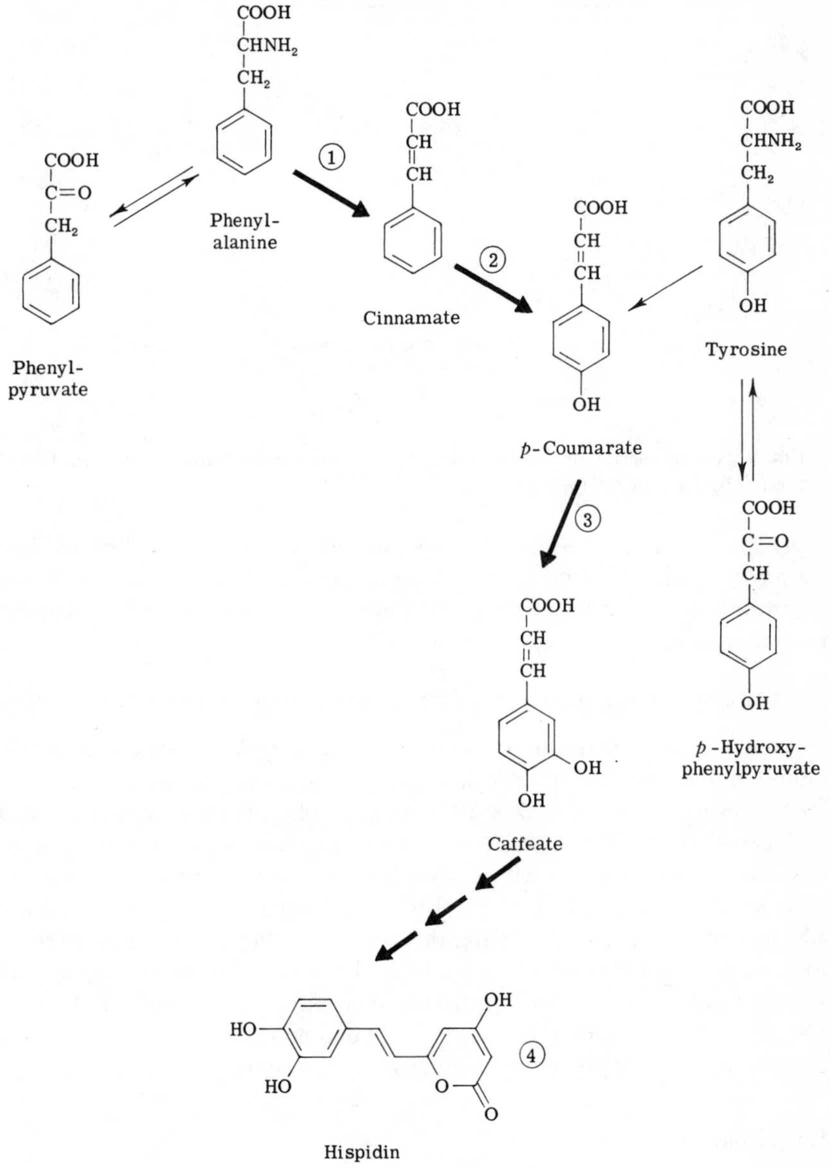

Fig. 8. Metabolic reactions studied in *Polyporus hispidus*. Heavy arrows indicate processes stimulated by light.

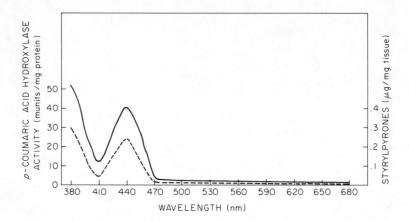

Fig. 9. Action spectra of styrylpyrone (– – –) and p-coumarate hydroxylase (——) induction by light in *Polyporus hispidus*.

biosynthesis and for the *in vivo* development of p-coumarate hydroxylase were very similar. The curves were bimodal with maxima at 380 and 440 nm and a minimum at 410 nm (Fig. 9). There was no stimulation of pigment formation or enzyme activity beyond 470 nm.

Effect of Cycloheximide and Chloramphenicol on the Light Effect

Even at a concentration as low as 25 μg/ml, cycloheximide completely inhibited p-coumarate hydroxylase and styrylpyrone formation in dark-grown cultures exposed to 5×10^5 ergs/cm², whereas chloramphenicol was ineffective at a concentration of 100 μg/ml. Inhibition of p-coumarate hydroxylase activity as well as styrylpyrone synthesis was dramatically influenced by the duration of the interval between the light pulse and the addition of cycloheximide. If the inhibitor was added immediately, there was complete inhibition; when added at later time intervals enzyme and styrylpyrone levels showed no further increases. Concentrations of up to 150 μg/ml of rifamycin or actinomycin D had no effect on either the levels of styrylpyrones or p-coumarate hydroxylase activity.

Conclusions

The light stimulation of enzymes of phenylpropanoid metabolism and of styrylpyrone biosynthesis in *Polyporus* is similar to the light stimulation of flavonoid biosynthesis, as well as the enzymes concerned, in cell suspensions of parsley (*Petroselinum hortense*) (Hahlbrock *et al.*, 1971; Hahlbrock, 1972).

There are important differences, however. Light triggers the appearance of PAL, cinnamate hydroxylase, and p-coumaryl 1:CoA ligase but not p-coumarate hydroxylase in the parsley system. In *Polyporus* PAL and cinnamate hydroxylase are affected by light, and p-coumarate hydroxylase activity shows a remarkable correlation with pigment production. In parsley, the qualitative or quantitative nature of the light requirement is not known. In *Helianthus tuberosus* the action spectrum for PAL development in illuminated disks shows a major peak at 440 nm and a minor one at 680 nm (Nitsch and Nitsch, 1966). Action spectra for enhanced PAL levels in etiolated barley shoots, determined at 1.1 and at 6.6 kerg/cm² showed peaks near 420, 620, and 660 nm (McClure, 1973). These action spectra further confirm phytochrome control of PAL in higher plants. The action spectrum of styrylpyrone biosynthesis in *Polyporus*, on the other hand, is similar to those determined for carotenoid biosynthesis in other fungi (Zalokar, 1955; Rillings, 1964). Other photomorphogenetic responses in fungi have also been associated with the blue region of the visible spectrum (Trione and Leach, 1969), and it has been suggested (Zolokar, 1955; Batra and Rillings, 1964; Rillings, 1964) that flavins may be the photoreceptors in these cases. Cinnamic acid 4-hydroxylase, one of the enzymes concerned with hispidin synthesis, requires FAD for maximum activity, but whether this is of any significance remains to be established.

Hartmann and Unser (1973) have presented arguments against flavins or carotenoids as the controlling pigments for blue-UV-mediated photoresponses, suggesting instead that the effector molecule is a phytochrome (P_{fr}). There are, to our knowledge, no suggestions in the literature of phytochrome involvement in light-induced phenomena in fungi.

What is the significance of styrylpyrone production in light and how common is the phenomenon of light-stimulated phenylpropanoid metabolism in Basidiomycetes? Hispidin production is reminiscent of anthocyanin and betacyanin production in higher plants. Perhaps the pigment is a UV screen which protects young, light-exposed mycelium in the same way that the young meristematic regions of some plants produce an abundance of anthocyanins. Careful studies of other Basidiomycetes may reveal a similar situation. We have found that there is a light effect on PAL production in other fungi. For example, cultures of *Sporobolomyces roseus*, grown in dark for five generations, produce twice as much PAL when these dark-grown cultures are treated with light. The same is true of other fungi such as *Clavaria*, although the phenomenon does not appear to be ubiquitous.

Our studies are at an early stage of development and we look forward to a better understanding of these photoregulated phenomena.

REFERENCES

Bandoni, R. J., K. Moore, P. V. Subba Rao, and G. H. N. Towers. 1968. *Phytochemistry* **7**:205–207.

Batra, P. P., and H. C. Rillings. 1964. *Arch. Biochem. Biophys.* **107**:485–492.

Brady, L. R., and R. G. Benedict. 1972. *J. Pharm. Sci.* **61**:318.

Bull, A. T., and B. L. A. Carter. 1973. *J. Gen. Microbiol.* **75**:61–73.

Bu'Lock, J. D., and H. G. Smith. 1961. *Experientia* **17**:553–554.

Bu'Lock, J. D. 1967. "Essays in Biosynthesis and Microbial Development." Wiley, New York.

Cain, R. B., R. F. Bilton, and J. A. Darrah. 1968. *Biochem. J.* **108**:797–828.

Camm, E. L., and G. H. N. Towers. 1969. *Phytochemistry* **8**:1407–1413.

Diamondstone, T. I. 1966. *Anal. Biochem.* **16**:395–401.

Edwards, R. L., D. G. Lewis, and D. V. Wilson. 1961. *J. Chem. Soc., London* pp. 4995–5002.

Hahlbrock, K. 1972. *FEBS Lett.* **28**:65–68.

Hahlbrock, K., I. Ebel, R. Ortman, A. Sutter, E. Wellmann, and H. Grisebach. 1971. *Biochim. Biophys. Acta* **224**:7–15.

Hartmann, K. M., and I. C. Unser. 1973. *Z. Pflanzenphysiol.* **69**:109–124.

Hatfield, G. M., and L. R. Brady. 1968. *Lloydia* **31**:225–228.

Hatfield, G. M., and L. R. Brady. 1969. *J. Pharm. Sci.* **58**:1298–1299.

Hatfield, G. M., and L. R. Brady. 1971. *Lloydia* **34**:260–263.

Hatfield, G. M., and L. R. Brady. 1973. *Lloydia* **36**:59–65.

McClure, J. 1973. *Phytochemistry* (in press).

Malstrom, B. G., and L. Ryden. 1968. *In* "Biological Oxidations" (T. P. Singer, ed.), pp. 415–438. (Wiley) Interscience, New York.

Moore, K., and G. H. N. Towers. 1967. *Can. J. Biochem.* **45**:1659–1665.

Moore, K., P. V. Subba Rao, and G. H. N. Towers. 1967. *Life Sci.* **6**:2629–2633.

Moore, K., P. V. Subba Rao, and G. H. N. Towers. 1968. *Biochem. J.* **106**:507–514.

Nambudiri, A. M. D., J. V. Bhat, and P. V. Subba Rao. 1972. *Biochem. J.* **130**:425–433.

Nambudiri, A. M. D., C. P. Vance, and G. H. N. Towers. 1973. *Biochem. J.* **134**:891–897.

Nitsch, C., and J. P. Nitsch. 1966. *C.R. Acad. Sci.* **262**:1102–1105.

Perrin, P. W. 1972. Ph.D. Dissertation. University of British Columbia.

Perrin, P. W., and G. H. N. Towers. 1973a. *Phytochemistry* **12**:583–587.

Perrin, P. W., and G. H. N. Towers. 1973b. *Phytochemistry* **12**:589–592.

Power, D. M., G. H. N. Towers, and A. C. Neish. 1965. *Can. J. Biochem.* **43**:1397–1407.

Rillings, H. C. 1964. *Biochim. Biophys. Acta* **79**:464–475.

Subba Rao, P. V., K. Moore, and G. H. N. Towers. 1967. *Can. J. Biochem.* **45**:1863–1872.

Subba Rao, P. V., B. Fritig, J. R. Vose, and G. H. N. Towers. 1971. *Phytochemistry* **10**:51–56.

Towers, G. H. N. 1969. *In* "Perspectives in Biochemistry" (J. B. Harborne and T. Swain, eds.), pp. 179–191. Academic Press, New York.

Towers, G. H. N., and P. V. Subba Rao. 1972. *Recent Advan. Phytochem.* **4**:1–43.

Trione, E. J., and C. M. Leach. 1969. *Phytopathology* **59**:1077–1083.

Ueno, A., S. Fukishima, Y. Saiki, and T. Harada. 1964. *Chem. Pharm. Bull.* **12**:376–378.

Vance, C. P., A. M. D. Nambudiri, and G. H. N. Towers. 1973. *Can. J. Biochem.* **51**:731–734.

Vaughn, P. F. T., V. S. Butt, H. Grisebach, and L. Schill. 1969. *Phytochemistry* **8**:1373–1378.

Walton, E., and V. S. Butt. 1970. *J. Exp. Bot.* **21**:887–891.

Zalokar, M. 1955. *Arch. Biochem. Biophys.* **56**:318–325.

NONPROTEIN AMINO ACIDS FROM PLANTS: DISTRIBUTION, BIOSYNTHESIS, AND ANALOG FUNCTIONS

LESLIE FOWDEN

Rothamsted Experimental Station, Harpenden, Hertfordshire, England

Introduction

DEFINITIONS AND STRUCTURAL TYPES

Nearly 25 years ago, I derived considerable satisfaction from a successful crystallization of γ-methyleneglutamine, obtained after fractionating the sap exuding from cut stems of young peanut plants. At that time, this third amino acid amide of plants, isolated a little earlier than the related dicarboxylic amino acid, γ-methyleneglutamic acid (Done and Fowden, 1952), was one of a very few unusual amino acids occasionally encountered in plants, coexisting with a normal complement of unbound forms of the protein amino acids. Gradually, as their number increased, these additional

95

compounds have commonly been described as "nonprotein" amino acids (Fowden, 1962), a name signifying their absence from protein molecules of the "producer" plant species. Now over 200 such amino acids (or simple, derived compounds such as γ-glutamyl peptides) have been characterized as plant constituents, and this number continues to grow as new species are examined, or as previously worked species are reexamined more critically.

Although the majority of compounds forming this group of plant constituents are small molecules having ten or fewer carbon atoms, a surprising variety of structural types within the group R [in $RCH(NH_2)CO_2H$] is represented in the group. For example, ethylenic unsaturation, first encountered in γ-methyleneglutamine, is now established as occurring in many other compounds, while the acetylenic link occurs as a characteristic feature of three C-7 amino acids isolated from seed of *Euphoria longan* (Sung *et al.*, 1969). A variety of heterocyclic ring systems (often attached as β-substituents on an alanine residue) is encountered and includes the pyridine, pyrazole, pyrimidine, furan, pyran, and isoxazoline groups. Other large groups of compounds include the S-substituted cysteines and the imino acids, based not only on the pyrrolidine nucleus typified by proline, but also upon the homologous azetidine and piperidine rings. Clearly, plants then show a fascinating versatility to create unorthodox nitrogenous compounds other than the more numerous and time-honored alkaloids. But, as a class, the nonprotein amino acids must be regarded, like the alkaloids, as secondary metabolic products, because very few of the known compounds have been allotted a function in the basic physiology or metabolism of those plants which produce them. The possibility that certain amino acids may be important in the sphere of physiological ecology has been discussed, especially with reference to the conferment of competitive advantage or protective value (Bell, 1972; Fowden, 1965).

DISTRIBUTION AND CONCENTRATION

Designation of a compound as a secondary product expresses a belief that the compound does not play an indispensable role within those plant species containing it, and that the compound need not occur generally in members of the plant kingdom. Indeed, this last concept provides the basis of the chemotaxonomic approach to plant classification, and in recent years the nonprotein amino acids have been featured prominently in this type of phytochemical study. Typically, plant extracts are routinely surveyed by applying chromatographic and electrophoretic techniques, when unusual amino acids representing principal constituents are recorded. Relationships at the level of species, genus, or family can then be examined against the distribution of characteristic associations of amino acids (often composed of the various intermediates of a particular biosynthetic pathway), rather than in relation to the occurrence of a single compound (see Bell and Fowden, 1964). For instance, the distribution of canavanine within the many genera forming the Papilionoideae does not provide a reliable taxonomic marker for distinguishing tribal groupings (Turner and Harborne, 1967), but the tribal affiliations of many individual genera of the Cucurbitaceae could be predicted from an examination of the occurrences of N^4-substituted asparagines (N-methyl, N-ethyl, and N-hydroxyethyl derivatives), m-carboxyphenylalanine, and β-(pyrazol-1-l)alanine and its γ-glutamyl peptide (Dunnill and Fowden, 1965). Nonprotein amino acids also have provided important guidelines in establishing subgeneric relationships within the following genera: *Lathyrus* (Bell, 1962), *Vicia* (Bell and Tirimanna, 1965), *Aesculus* (Fowden *et al.*, 1970), *Acacia* (Seneviratne and Fowden, 1968), *Astragalus* (Dunnill and Fowden, 1967), and *Acer* (Fowden and Pratt, 1973).

It is important to stress that the successes achieved by employing the chemotaxonomic approach have depended upon the *accumulation* of secondary compounds to levels that are easily recognized by routine analytical techniques. A noteworthy presence of a compound then has more significance than an absence, since a record of the latter type more precisely indicates that the compound was not elaborated in amounts detectable by the techniques employed. Here, I do not wish to be fussy but merely to focus attention upon certain genetical considerations underlying secondary product biosynthesis. Does a chemotaxonomic approach based upon amino acid distributions necessarily depend upon the fact that plants either possess or lack totally genes eliciting particular biosynthetic reactions? Or do such genes have an almost universal distribution among members of the plant kingdom? Different patterns of amino acid accumulation (from trace amounts escaping routine detection to the massive accumulations associated with some species) would then reflect differences in the degree to which particular genes were "switched on."

Quite recently, I had a unique experience that was to provide pertinent comment upon this problem. I was invited to collaborate with Dr. H. Knobloch (Paris) in the identification of certain less-common amino and imino compounds resulting from large-scale, commercial separations of the nitrogenous fraction present in sugar beet extracts destined for sugar refining (see Fowden, 1972a). The fractionation techniques employed included cation- and anion-exchange resin chromatography, followed by crystallization—namely, those normally adopted in laboratory practice—but the scale of working was multiplied at least 100,000-fold. A batch fractionation might commence typically with 50,000 kg of mixed nitrogenous compounds (a cationic fraction), originating from about 10^5 ton of fresh beet. Analysis of individual fractions revealed the presence of very many "unidentified" ninhydrin-positive compounds, most of which remain uncharacterized. However, the isolation of about 30 kg of azetidine-2-carboxylic acid (A2C) was of particular interest because there was no previous record of its occurrence in sugar beet or in any other member of the family Chenopodiaceae. The imino acid is a constituent of a number of liliaceous species (Fowden and Steward, 1957) and of a very few legumes (Sung and Fowden, 1969; Watson and Fowden, 1973). Although A2C was isolated in substantial quantity, it is important to realize that its concentration in fresh beet was about 0.3 ppm, i.e., a level well below the threshold concentration that would be detected by routine chromatographic methods. Sugar beet then provides an instance of a plant having the genetic potential for A2C biosynthesis, but clearly either the genes controlling the synthesis of the biosynthetic enzymes are strongly repressed or the catalytic activity of certain enzymes is markedly inhibited under normal growth conditions. Two further unusual imino acids were present in beet at slightly higher concentrations: these were pipecolic acid (encountered sporadically in plants) and its unsaturated derivative, baikiain (4,5-dehydropipecolic acid), previously isolated only from legumes (*Baikiaea plurijuga*, King *et al.*, 1950; *Caesalpinia tinctoria*, Watson and Fowden, 1973). A number of *N*-acylated diamino acids present in the mother liquors remaining after crystallization of glutamic acid were new isolates from plants (Fowden, 1972a). This work then strongly suggests that many compounds now regarded as having a restricted distribution in plants are in fact elaborated by a much wider range of plants, but only in amounts that escape detection by normal laboratory procedures.

Phenylalanine Derivatives, Homologs, and Analogs

In this section, a more detailed treatment of the occurrence, biosynthesis, and analog roles of the nonprotein amino acids will be developed, but of necessity only a few compounds will be surveyed. One group of compounds that illustrates features of more general interest consists of phenylalanine and its substituted and homologous derivatives.

The structures of some of these compounds are given in Table 1, together with their sources of origin, which are largely angiosperms. Optical configurations are included where known or suggested by application of the Lutz-Jirgensons rule after the determination of specific rotations in water and mineral acid: certain substituted phenylglycines are definitely D-isomers, while *N*-carbamoyl-*p*-hydroxyphenylglycine was isolated as a racemic form. Compounds **6, 7, 11,** and **14** have been characterized since these aromatic amino acids were last reviewed (Fowden, 1970); **6** and **7** occur as important constituents of seed of *C. tinctoria*, but the compounds could not be detected in seed of 12 other *Caesalpinia* spp. examined (Watson and Fowden, 1973). Similarly, 4-aminophenylalanine was detected only in *Vigna vexillata* and *V. nuda*, although a large number of other members of the Phaseolinae were screened (Dardenne *et al.*, 1972).

BIOSYNTHESIS

Aromatic hydroxylation has been the subject of many original studies and review articles, and so the present account will be confined entirely to considerations involving compounds listed in Table 1. If tyrosine is formed in plant systems by direct hydroxylation of phenylalanine [the claim by Nair and Vining (1965) to have demonstrated this conversion has been strongly contested by a number of other workers], the reaction is quantitatively far less important than the alternative route involving amination of *p*-hydroxyphenylpyruvic acid. Similarly, Müller and Schütte (1967) could find no support for a direct conversion of ^{14}C-phenylalanine into *m*-tyrosine when the labeled amino acid was fed to plants of *Euphorbia myrsinites*. In contrast, feeding of ^{14}C-shikimate led to an incorporation of label into *m*-tyrosine, and the German workers believed that the most probable biosynthetic route involved an early hydroxylation reaction occurring before the aromatization of the ring system, i.e., a variant of the normal shikimate pathway was implicated (cf. the biosynthesis of *m*-carboxy-substituted phenylalanines below). It is possible that *m*- and *p*-hydroxyphenylglycines arise from *m*-tyrosine and tyrosine, respectively, by reactions analogous to those effecting side-chain reduction leading to the formation of *m*-carboxy-substituted phenylglycines from the corresponding alanine derivatives (see below).

The single C substituents attached to the phenyl ring, and existing at various oxidation levels (methyl, hydroxymethyl, and carboxyl) in the different compounds, originate in a variety of ways. The CH_3 group of orcylalanine (**5**) arises directly from acetate: among the amino acids in Table 1, **5** seems unique, being the only compound in which the aromatic ring arises by acetate (or malonyl-CoA) polymerization. The *o*- and *p*-hydroxy groups of orcylalanine are also inserted directly during polymerization, which probably initially yields orsellinic acid; condensation (and simultaneous decarboxylation) of orsellinic acid with a C_3 moiety of serine is the most likely pathway producing orcylalanine in *Agrostemma* plants (Hadwiger *et al.*, 1965). In constrast, *p*-hydroxymethylphenylalanine apparently is synthesized in *E. coli* from phenylalanine, presumably by a process adding a C_1 unit at the *para* position (Sloane and Smith, 1968).

TABLE 1

SOME NATURALLY OCCURRING AROMATIC β-SUBSTITUTED ALANINES [RCH₂CH(NH₂)CO₂H] AND α-SUBSTITUTED GLYCINES [RCH(NH₂)CO₂H]

TABLE 1

SOME NATURALLY OCCURRING AROMATIC β-SUBSTITUTED ALANINES $[RCH_2CH(NH_2)CO_2H]$ AND α-SUBSTITUTED GLYCINES $[RCH(NH_2)CO_2H]$

Structure No.	Compound		Occurrence	Reference
β-Substituted Alanines				
1	L-Phenylalanine	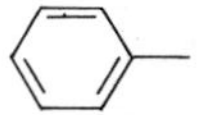	Universally	
2	L-Tyrosine	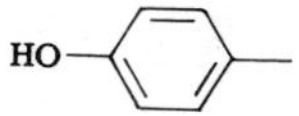	Universally	
3	L-*m*-Tyrosine	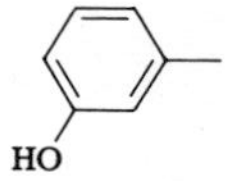	*Euphorbia myrinitis*	Mothes *et al.* (1964)
4	L-3,4-Dihydroxyphenylalanine (dopa)	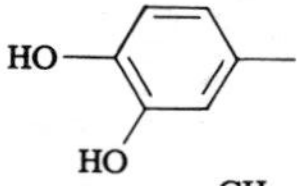	*Vicia faba, Mucuna* spp.	Torquati (1913) Bell and Janzen (1971) Daxenbichler *et al.* (1971)
5	β-Orcylalanine	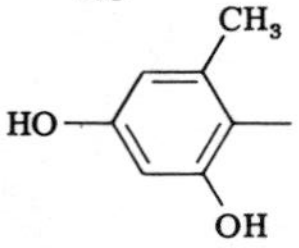	*Euphorbia* spp., *Agrostenema githago*	Liss (1961); Adinolfi (1964) Schneider (1958)
6	3-Hydroxymethylphenylalanine	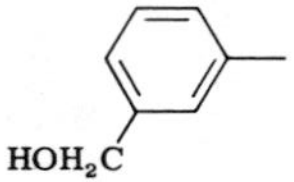	*Caesalpinia tinctoria*	Watson and Fowden (1973)
7	L-4-Hydroxy-3-hydroxymethyl-phenylalanine	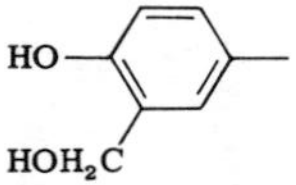	*Caesalpinia tinctoria*	Watson and Fowden (1973)

α-Substituted Glycines

No.	Compound	Source	Reference
8	4-Hydroxymethylphenylalanine	*Escherichia coli*	Sloane and Smith (1968)
9	L-*m*-Carboxyphenylalanine	*Iridaceae, Resedaceae, Cucurbitaceae C. tinctoria*	Thompson *et al.* (1961) Kjaer and Larsen (1963) Dunhill and Fowden (1965) Watson and Fowden (1973)
10	L-*m*-Carboxytyrosine	*Resedaceae*	Kjaer and Larsen (1963)
11	L-4-Aminophenylalanine	*Vigna vexillata*	Dardenne *et al.* (1972)
12	Phenylglycine	Phloem sap from *Fagus*	Dietrichs and Funke (1967)
13	*m*-Hydroxyphenylglycine	*Euphorbia helioscopia*	Müller and Schütte (1967)
14	*N*-Carbamoyl-DL-*p*-hydroxy-phenylglycine	*Vicia faba* leaf	Eagles *et al.* (1971)
15	3,5-Dihydroxyphenylglycine	*Euphorbia helioscopia*	Müller and Schütte (1967)
16	D-*m*-Carboxyphenylglycine	*Resedaceae, Iridaceae*	Kjaer and Larsen (1963) Morris *et al.* (1959)
17	D-3-Carboxy-4-hydroxyphenyl-glycine	*Reseda* spp.	Kjaer and Larsen (1963)

While similar C_1 transfer mechanisms could be conceived to explain the biosynthesis of compounds **6, 7, 9** and **10,** which possess either a hydroxymethyl or carboxyl group at the *m*-position, the use of stereospecifically labeled shikimic acids has established beyond doubt that the *m*-carboxy derivatives are produced from shikimate by steps involving a rearrangement mechanism different from that encountered in the biosynthesis of phenylalanine and tyrosine. In this alternative process, the carboxyl group of the original shikimate molecule is retained as the *m*-substituent in the final ring-carboxylated amino acids. This was established conclusively in experiments in which (carboxy-^{14}C) shikimate was supplied to plants of *Reseda odorata* and *R. lutea,* label introduced into *m*-carboxyphenylalanine and *m*-carboxytyrosine was located almost exclusively in the *m*-carboxyl-C atom (Larsen, 1967).

Larsen *et al.* (1972) have recently published a more detailed study of the steric course and rearrangements occurring during the conversion of shikimic acid into phenylalanine, tyrosine, *m*-carboxyphenylalanine, and *m*-carboxytyrosine in *R. lutea* plants. For these experiments they synthesized doubly labeled shikimic acid molecules possessing (U-^{14}C) label and specific tritium labeling either at the 6R- or 6S-positions. Figure 1 illustrates the reaction scheme adopted to explain their observations relating to the retention or loss of tritium label when the different amino acids were produced from 6R- and 6S-forms of ^{3}H-shikimate (**18**). The 6S-hydrogen is retained and the 6R-hydrogen is lost during enzymatic transformation of **18** into **20** by *Aerobacter aerogenes* and *E. coli* extracts (Onderka and Floss, 1969; Floss *et al.*, 1972; Hill and Newkome, 1969), and the new studies with *Reseda* indicate a similar complete retention of the 6S-hydrogen of shikimate during biosynthesis of phenylalanine (**1**), tyrosine (**2**), and *m*-carboxyphenylalanine (**9**). The location of tritium in the *ortho* position in phenylalanine (**1**) and in tyrosine (**2**) is that expected when the C_3 side chain becomes attached to the ring C-1 atom of 18: thus direct evidence for the rearrangement of chorismic acid (**20**) into prephenic acid (**21**) in higher plants was obtained for the first time. The observed retention of the 6S-hydrogen of shikimate in *m*-carboxyphenylalanine (**9**) supported the view that this amino acid is also derived by conversion of shikimate (**18**) into chorismate (**20**). The location of tritium at the *para* position in **9** demonstrated that the C_3 side chain migrates to the ring C-3 atom of the original shikimate. A probable pathway would involve sequential rearrangements of chorismate (**20**) into isochorismic acid (**22**) and isoprephenate (**23**), followed by elimation to give *m*-carboxyphenylpyruvic acid and its transamination to *m*-carboxyphenylalanine (**9**). These rearrangements are thermally permissable processes.

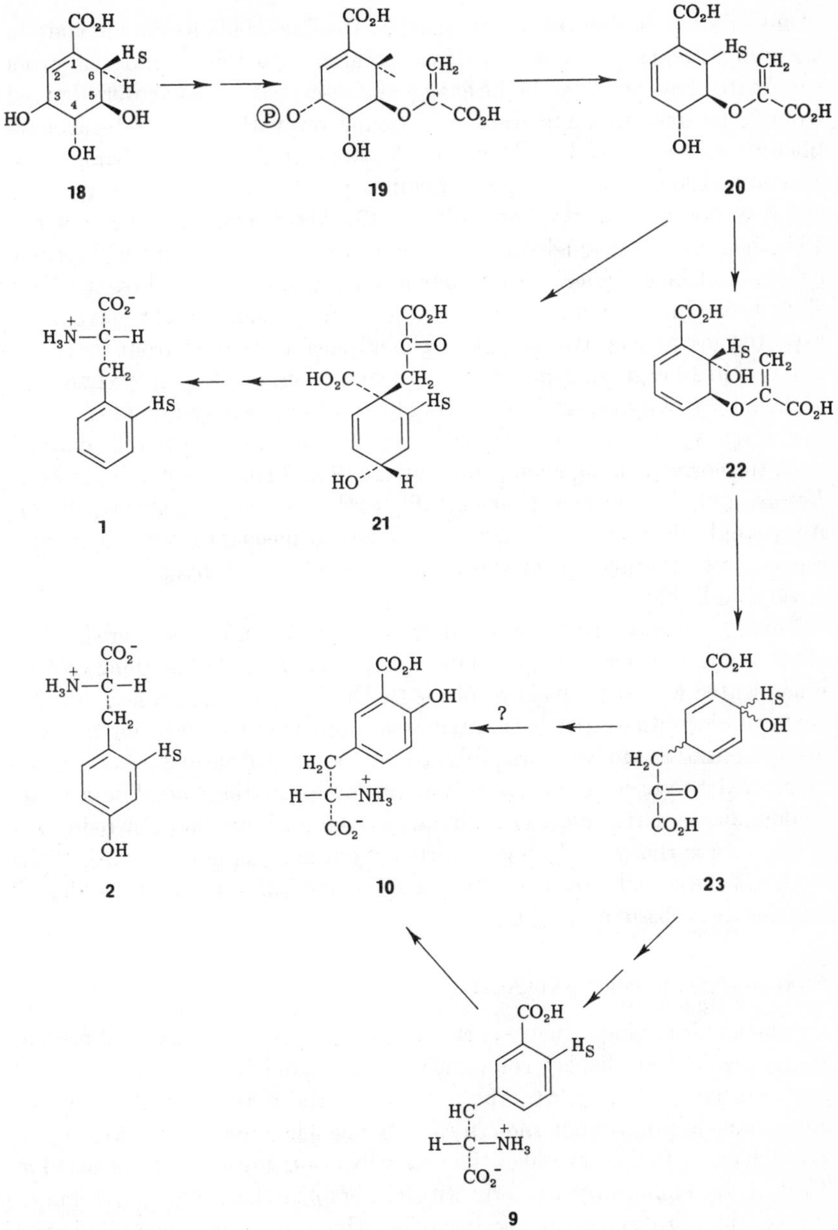

FIG. 1. Scheme illustrating stereospecific biosynthesis of *m*-carboxy-substituted aromatic amino acids from labeled shikimic acid (after *Larsen et al.*, 1972). Compounds are numbered as in text.

m-Carboxytyrosine (**10**) synthesized by the *Reseda* plants did not contain tritium, although ^{14}C-label was incorporated into this amino acid from shikimate. This result would be expected if *m*-carboxytyrosine were derived from **23** by oxidation and transamination. However, when ^{14}C-*m*-carboxyphenylalanine was fed to leaves of *R. lutea* or *R. odorata* (Larsen and Sorensen, 1968), low ^{14}C-activity was incorporated into *m*-carboxytyrosine and 3-carboxy-4-hydroxyphenylglycine (**17**). Direct hydroxylation of **9** into **10** is then feasible (though an intermediary role of *m*-carboxyphenylpyruvic acid cannot be excluded). If hydroxylation were accompanied by the NIH shift (see Guroff *et al.*, 1967), then tritium labeling might have been expected in the ring C-5 position. The absence of tritium from this *ortho* position in **10** argues against direct hydroxylation, but loss by exchange to a phenolic group during isolation could not be completely excluded.

m-Carboxy-substituted phenylglycines (**16** and **17**) appear to be derived from the corresponding alanine derivatives (**9** and **10**, respectively) in both *Reseda* and *Iris* species (Larsen, 1967; Morris and Thompson, 1965). Additional information relating to the reaction mechanisms involved, and especially to the manner in which D-isomers of the phenylglycines result, is very desirable.

Finally, the biosynthesis of compounds **6** and **7** can be considered. They are present in moderate amounts in seed of *C. tinctoria*, but seedling growth is associated with a gradual loss of the two hydroxymethyl amino acids. No study of biosynthesis has been attempted, but the use of maturing fruits as experimental tissue will probably be necessary to ensure success. The associated presence of *m*-carboxyphenylalanine in the *Caesalpinia* seeds could indicate that 3-hydroxymethylphenylalanine is produced by reduction of the ring carboxyl of **9**, but *m*-carboxytyrosine was not detected in the seeds (Watson and Fowden, 1973). Direct hydroxylation of **6** to yield **7** remains as a possible reaction.

ANALOGS: STRUCTURE AND ACTIVITY

Substances having structures closely akin to those of individual protein amino acids often display competitive or antagonistic activity in normal processes concerned with (i) the uptake, (ii) the biosynthesis, or (iii) the incorporation into protein molecules of particular amino acids (see review, Fowden *et al.*, 1967). Several of the naturally occurring amino acids listed in Table 1 are sufficiently isosteric with either phenylalanine or tyrosine to enable them to show analog behavior which may be most obviously demonstrated as a general inhibition of bacterial (see Tristram, 1973) or seedling growth (see Fowden, 1963). Other natural amino acids, including compounds based on structure **24** and β-(pyrazol-1-yl)alanine (**25**) will be considered in this section, as will the phosphonic acid (26) and tetrazole (27) derivatives of phenylalanine.

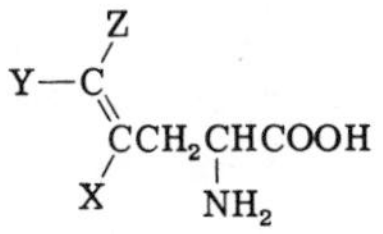

24

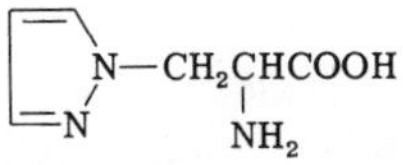

25

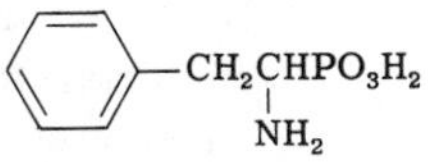

26

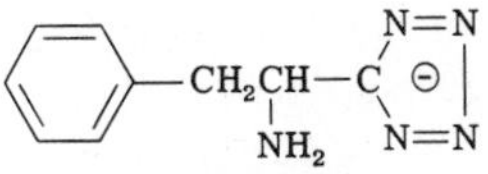

27

Permease Studies with Bacterial Systems

Several investigators have studied the properties of the permease enzyme systems responsible for the transport of aromatic amino acids across the cell membrane. Bacterial studies, especially with strains of *E. coli* or *Salmonella typhimurium*, have indicated that cells commonly possess a general aromatic amino acid permease which displays a broad specificity and transports phenylalanine, tyrosine, and tryptophan as well as a wide variety of aromatic amino acid analogs. In addition, these organisms possess three other aromatic permeases, each specific for a single amino acid, i.e., phenylalanine, tyrosine, and tryptophan, respectively (Ames, 1964; Ames and Roth, 1968; Brown, 1970; Willshaw and Tristram, 1972). It is possible to distinguish between the action of the general and specific permeases because only the former type is functional when amino acid incorporation into protein is inhibited by the presence of chloramphenicol (150 μg/ml) during uptake (Willshaw and Tristram, 1972.)

In general, it is necessary for an analog to possess an unmodified alanine moiety if it is to show significant affinity for the general aromatic permease. Analogs not transported then include compounds in which a glycine or a β-hydroxyalanine moiety replaced the normal alanine side chain, or others in which the carboxyl group was replaced by a phosphonic acid group (26) or a tetrazole ring (27). Analogs based on ring-substituted phenylalanines were frequently good substrates; a single small substituent group such as –F (in *p*-fluorophenylalanine) led to a far smaller alteration in the affinity of the analog for the permease than was caused by substitution of large (e.g., *p*-CH$_3$O) or more polar (e.g., *m*-CO$_2$H) groups. A number of compounds in which the phenyl ring was replaced by other aromatic systems (as in β-pyrazol-1-ylalanine, β-thien-2-ylalanine, or β-pyridin-2-ylalanine) also behaved as effective analogs and showed significant affinity for the general aromatic amino acid permease.

The analog behavior, in relation to the aromatic permeases, of compounds based on structure 24 is illustrated in Table 2. 2-Amino-4-methylhex-4-enoic acid (24a, where $X = Y = CH_3$ and $Z = H$) represents the major free amino acid of seed of *Aesculus californica* (Smith and Fowden, 1968), while 2-amino-4-ethylpent-4-enoic acid (24b, where $X = C_2H_5$ and $Y = Z = H$) is a chemically synthesized isomer. These compounds exhibit a structural similarity to phenylalanine, since the possession of a double bond at position-4,5 allows coplanarity of the adjacent C atoms; as a result the permease enzymes accept these molecules, recognizing them by the intactness of the alanine moiety and by the presence of a smaller (or incomplete) phenyl ring. Quantitative measurements of the affinities of analogs for the permeases were obtained by assaying the degree of inhibition of accumulation of ^{14}C-labeled phenylalanine or tyrosine, because labeled analogs were not generally available. Compound 24a exhibited higher affinities than 24b for both the general aromatic permease and for the

TABLE 2

Analog Inhibition[a] of Accumulation of ^{14}C-Phenylalanine and ^{14}C-Tyrosine by Cells of *E. coli* and of the Phenylalanine-Sensitive DAHP Synthetase[b]

Inhibitor	Phe-specific permease[a]	Tyr-specific permease[c]	General aromatic permease[d] Phe	General aromatic permease[d] Tyr	Phe-sensitive DAHP synthetase[e]
L-Phenylalanine	95	20	—	—	94
L-Tyrosine	38	92	—	—	12
DL-2-Amino-4-methylhex-4-enoic acid	95	13	67	65	47
DL-2-Amino-4-ethylpent-4-enoic acid	82	22	32	11	68
L-β-Pyrazol-1-ylalanine	26	0	23	—	82

[a] Inhibition expressed as a % of initial rate in absence of inhibitors.

[b] After Tristram (1973).

[c] Accumulation of ^{14}C-phe or ^{14}C-tyr (at 2×10^{-6} M) measured in presence (or absence) of inhibitors at 2×10^{-4} M conc.

[d] Accumulation of ^{14}C-phe or ^{14}C-tyr (at 2×10^{-7} M) measured in presence (or absence) of inhibitors at 5×10^{-5} M conc.

[e] Inhibitor conc. 10^{-3} M.

phenylalanine-specific enzyme: both compounds acted as strong antagonists of the phenylalanine-specific permease, but they showed only weak inhibition of the tyrosine-specific enzyme. Saturation of the double bond in **24a** would cause loss of planarity about C-4 and gave a compound (2-amino-4-methylhex-4-anoic acid, a homoisoleucine), showing only low affinity for these aromatic permeases.

Analog Effects upon Amino Acid Uptake by Plant Roots

R. Watson and L. Fowden (unpublished experiments) have begun a study designed to provide similar information concerning the effect of structural analogs of phenylalanine and/or tyrosine upon the uptake of the two protein amino acids by roots of young seedlings. Particular attention was focused upon the possible behavior of the two "new" aromatic amino acids, 3-hydroxymethylphenylalanine (**6**) and 3-hydroxymethyl-4-hydroxyphenylalanine (**7**), isolated from seed of *Caesalpinia tinctoria*. Uptake of labeled phenylalanine or tyrosine by young roots of *Caesalpinia* or melon (*Cucumis melo*) seedlings was studied over a period of an hour; after a rapid uptake of labeled amino acid in the first few minutes (due to adsorption onto root surfaces or the diffusive filling of "free space"), the rate of further uptake remained constant during the experiment. This later active uptake, representing 70–75 percent of the total uptake occurring in an hour, was almost totally eliminated by the presence of respiratory inhibitors such as cyanide or azide. When uptake studies were repeated in the presence of analog molecules (used at concentrations 50× higher than those of the labeled phenylalanine or tyrosine), clear evidence of competition for sites on the permease was obtained for several analogs (see Table 3). Since these are unlikely to affect diffusive uptake, inhibitions of 70–75 percent of the total uptake recorded with some analogs probably indicate an almost total suppression of the normal "active" uptake process. The experimental data suggest that products **6** and **7**, unique to *Caesalpinia*, antagonize the uptake of phenylalanine and tyrosine to a lesser extent in this species than in a nonproducer species such as melon. L-Dopa displays a similar effectiveness as a permease competitor, but introduction of an α-methyl group is associated with a marked reduction in analog activity.

Analogs as Feedback Inhibitors of Aromatic Amino Acid Biosynthesis

Studies of this type have been confined to microbial systems, with emphasis having been placed upon inhibitory effects on 3-deoxy-D-*arabino*-heptulosonic acid 7-phosphate (DAHP) synthetase, the enzyme catalyzing the first distinctive reaction of aromatic biosynthesis. Normally,

TABLE 3

Inhibition of Uptake of l-Phenylalanine (50 μM) and l-Tyrosine (50 μM) by
Various Analogs (2.5 mM) in Young Seedlings of *Caesalpinia tinctoria*
and *Cucumis melo*[a]

Analog	Inhibition of phe uptake		Inhibition of tyr uptake	
	Caesalpinia	*Cucumis*	*Caesalpinia*	*Cucumis*
Phenylalanine	—	—	60	60
Tyrosine	57	62	—	—
3,4-Dihydroxyphenylalanine-(dopa)	55	67	57	69
α-Methyldopa	6	13	13	52
3-Hydroxymethylphenylala-nine	44	56	36	73
3-Hydroxymethyl-4-hydroxy-phenylalanine	49	50	35	74

[a] Data expressed as % inhibition of normal uptake in absence of an analog.

three isoenzymic forms of DHP synthetase coexist in cells and show, respectively, sensitivity to phenylalanine, tyrosine, and tryptophan. Smith *et al.* (1962) have tested a large number of phenylalanine and tyrosine analogs for ability to inhibit the phe- and tyr-sensitive forms of the enzyme from *E. coli* W. Some ring-substituted fluoro or methyl derivatives of phenylalanine were as effective as phenylalanine itself in inhibiting the activity of phe-sensitive DAHP synthetase; β-thien-2-(or 3-)ylalanines were also strong inhibitors. The tyr-sensitive DAHP synthetase showed a stricter stereospecificity and only analogs in which the hydroxyl group at ring C-4 was preserved exhibited marked inhibitory action, e.g., 3-fluorotyrosine and 3,4-dihydroxyphenylalanine.

A later study has utilized β-pyrazol-1-ylalanine (**25**) and open-chain analogs (type structure **24**) in a similar study of the phe-sensitive DAHP synthetase of an *E. coli* strain lacking tyr- and trp-sensitive forms of the enzyme (see Tristram, 1973). The specificity of the allosteric site of the DAHP synthetase differed markedly from that of the amino acid binding site of the general aromatic permease, e.g., 10^{-3} M β-pyrazol-1-ylalanine caused 82 percent inhibition of synthetase activity, wherease the compound was poorly transported by the general aromatic amino acid permease (see Table 2). In contrast, 2-amino-4-methylhex-4-enoic acid showed high affinities for both the general aromatic and phe-specific permease but was less effective as an inhibitor of phe-sensitive DAHP synthetase (cf. also Smith *et al.*, 1962). The isomeric 2-amino-4-ethylpent-4-enoic acid shows the reverse behavior, displaying a marked inhibitory action against the synthetase, a high affinity for the phe-specific permease, but little affinity for the general aromatic amino acid permease.

Experiments of this type have demonstrated that analogs can affect the rate of synthesis of at least some amino acid constituents of protein, and thereby tend to limit the rate of formation of protein molecules themselves. However, we completely lack information concerning the possibility that the production of unusual aromatic amino acids by plants may be controlled by similar feedback inhibition (or repression) of key biosynthetic enzymes: such uncertainties clearly delineate areas requiring detailed study in the future.

Analogs in Relation to Aminoacyl-tRNA Synthetases and Protein Synthesis

Many investigations have shown that modified phenylalanines (especially ring-substituted fluoro derivatives) interfere with protein synthesis in a whole range of organisms. Antagonistic effects arise by competitive binding of the analog at the active sites of the aminoacyl-tRNA synthetases, thereby causing (i) an inhibition of the rate of protein synthesis by limiting the availability of appropriate aminoacyl residues or (ii) the incorporation of analogs into protein molecules with consequent modification of molecular conformation, or both (i) and (ii) simultaneously. A more detailed consideration of the mechanisms whereby analogs may produce inhibitions of growth and differentiation, and of metabolic processes is available in two reviews (Fowden *et al.*, 1967; Fowden, 1972b).

In our own work, we have attempted to define the range of molecular structures compatible with binding at the active site of phenylalanyl-tRNA synthetase, and to a lesser extent of tyrosyl-tRNA synthetase, from a variety of higher plant systems, and to study possible variations in the substrate specificity of a particular synthetase when obtained from different plant species. Earlier studies with phenylalanyl-tRNA synthetase from strains of *E. coli* had indicated that the specificity of this enzyme was not exacting and that it would activate a variety of analogs of phenylalanine (Conway *et al.*, 1962). Furthermore, the substrate specificity towards analogs was altered in mutants selected for resistance to growth inhibition towards analogs (such as *p*-fluorophenylalanine) which normally produced strong inhibition of wild-type cultures (Fangman and Neidhardt, 1964); the phenylalanyl-tRNA synthetase from a mutant resistant to *p*-fluorophenylalanine showed a negligible affinity for the analog, although it still strongly activated the *o*- and *m*-fluoro derivatives. Our studies endorse the bacterial findings and we conclude that the amino acid substrate specificity, especially of phenylalanyl-tRNA synthetase, is particularly loose.

TABLE 4

THE ABILITY OF VARIOUS ANALOGS TO STIMULATE
ATP-^{32}PP$_i$ EXCHANGE CATALYZED BY
PHENYLALANYL-tRNA SYNTHETASE
FROM *P. aureus*

Amino acid substrate[a]	Rate of ATP-^{32}PP$_i$ exchange (cf. phenylalanine = 100)
p-Fluorophenylalanine	63
o-Fluorophenylalanine	61
Tyrosine	8
m-Tyrosine	27
3,4-Dihydroxyphenylalanine	0
2-Amino-4-methylhex-4-enoic acid	114
2-Amino-4-methylhex-4-anoic acid	3

[a] Present at 5 mM conc. in assay mixtures, except tyrosine, which was used as a saturated solution.

Phenylalanyl-tRNA synthetase was purified initially from seed of *Phaseolus aureus* (Smith and Fowden, 1968). The enzyme's ability to activate a range of analogs, in comparison with the normal substrate phenylalanine, is illustrated in Table 4: activities are expressed in terms of an analog's effectiveness in stimulating the ATP-^{32}PP$_i$ exchange reaction catalyzed by phenylalanyl-tRNA synthetase. The presence of a single fluoro or hydroxyl substituent on the ring is compatible with a retention of substrate activity. Even the small activation recorded for tyrosine was apparently real, and not due to contaminant tyrosyl-tRNA synthetase. However, the most striking feature of these experiments was the very high level of substrate activity recorded for 2-amino-4-methylhex-4-enoic acid (**24a**). Apparently, a completed phenyl ring is not an obligatory feature of a substrate, but coplanarity of the C atoms surrounding C-4 is obviously essential, because the saturated homoisoleucine shows negligible substrate activity. The specificity of tyrosyl-tRNA synthetase was more exacting, and few analogs showed measurable substrate activity; however, 3-fluorotyrosine gave a V_{max} value approximately 30 percent of that measured for tyrosine.

The ability of phenylalanyl-tRNA synthetase to utilize open-chain amino acids of structure type **24** as substrates was further tested using enzyme preparations from species of *Aesculus*. *A. californica* (assigned to the subgeneric section, Calothyrsus) synthesizes large amounts of 2-amino-4-methylhex-4-enoic acid, and minor quantities of 2-amino-4-methylhex-4-anoic acid and 2-amino-6-hydroxy-4-methylhex-4-enoic acid (Fowden and and Smith, 1968), but these compounds are not detected in *Aesculus* species placed in the four other subgeneric groups. The ability of each compound to stimulate ATP-^{32}PP$_i$ exchange catalyzed by phenylalanyl-tRNA synthetase originating from *A. hippocastanum* is shown in Table 5A (Anderson and Fowden, 1970a). Two other compounds from among this group of un-saturated open-chain amino acids are natural products, namely, 2-amino-hex-4,5-dienoic acid (from *Amanita solitaria*, Chilton *et al.*, 1968) and 2-amino-5-methylhex-4-enoic   acid   (from   *Leucocortinarius bulbiger*, Dardenne *et al.*, 1968). The affinity of individual molecules for binding at the active site of the enzyme increased as C atoms replaced H at positions X and Y in structure **24,** and in crude terms when a phenyl ring had been reduced to two carbon atoms, i.e., allylglycine, all substrate activity was lost. Again 2-amino-4-methylhex-4-enoic acid (**24a**) had a V_{max} equal to that of the normal substrate phenylalanine, but the K_m for **24a** was about $1.2 \times 10^{-3}\ M$, whereas the value for phenylalanine was $3 \times 10^{-5}\ M$.

The enzyme from *A. hippocastanum* was employed with three other groups of phenylalanine analogs. In the first of these groups, various substituents were introduced into the phenyl ring. Activity data are given in Table 5B. The presence of large polar or ionizable groups is incompatible with affinity for the enzyme, and tyrosine is completely inert with this enzyme (contrast the situation with *P. aureus* enzyme above). The second group consisted of amino acids based on an alanine moiety substituted at the β-position by a variety of heterocyclic rings which replaced the normal phenyl ring of phenylalanine. Thienyl, pyridinyl, and pyrazol-1-yl deriva-tives were accepted as substrates by the synthetase (Table 5C), and so the same structural features are compatible with enzyme affinity as were found for the general aromatic amino acid permease. The final group was composed of amino acids containing modified alanine side chains attached to an unaltered phenyl ring. Longer or shorter side chains were incompatible with substrate activity, but when a β-hydroxyl group (as in β-phenylserine) was introduced, the molecule retained considerable enzyme affinity.

Phenylalanyl-tRNA synthetases have been purified from seed of five species of *Aesculus* (a representative from each of the recognized subgeneric groups) and the affinities of these enzymes for substrates, especially towards phenylalanine and 2-amino-4-methylhex-4-enoic acid, have been compared (Anderson and Fowden, 1970a). All five enzymes had very similar K_m values for ATP; more variation was encountered between K_m values measured for

TABLE 5

RELATION BETWEEN STRUCTURE AND ACTIVITY OF COMPOUNDS USED AS SUBSTRATES FOR THE PHENYLALANYL-tRNA SYNTHETASE OF *A. hippocastanum*[a]

			Compound[b]	$\dfrac{V_{max} \text{ analog}}{V_{max} \text{ phe}} \times 100$

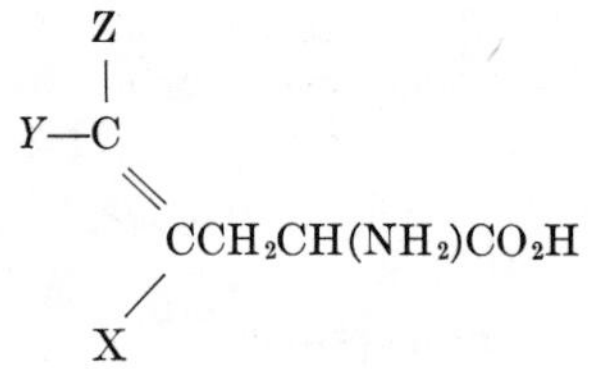

(A)

$$\begin{array}{c} Z \\ | \\ Y{-}C \\ \diagdown \\ \qquad CCH_2CH(NH_2)CO_2H \\ \diagup \\ X \end{array}$$

	X =	Y =	Z =		
(i)	H	H	H	Allylglycine (5 mM)	0
(ii)	H		CH₂	2-Aminohex-4,5-dienoic acid (5 mM)	23
(iii)	H	CH₃	H	Crotylglycine (2 5 mM)	29
(iv)	CH₃	H	H	Methallylglycine (5 mM)	35
(v)	C₂H₅	H	H	Ethallylglycine (5 mM)	40
(vi)	H	CH₃	CH₃	2-Amino-5-methylhex-4-enoic acid (10 mM)	48
(vii)	CH₃	CH₃	H	2-Amino-4-methylhex-4-enoic acid (10 mM)	100

(B)

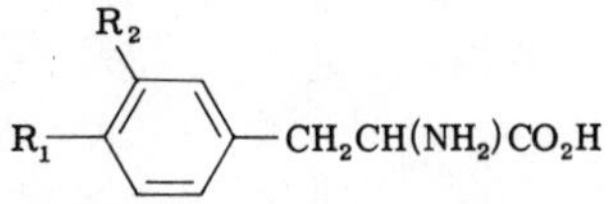

$$\begin{array}{c} R_2 \\ | \\ R_1{-}{-}CH_2CH(NH_2)CO_2H \end{array}$$

	R₁ =	R₂ =		
(i)	F	H	*m*-Fluorophenylalanine (2.5 mM)	52
(ii)	CO₂H	H	*m*-Carboxyphenylalanine (5 mM)	0
(iii)	H	OCH₃	*p*-Methoxyphenylalanine (2 5 mM)	42
(iv)	H	OH	Tyrosine (5 mM)	0
(v)	CN	H	*m*-Cyanophenylalanine (5 mM)	0

TABLE 5—(Continued)

Compound[b]	$\dfrac{V_{max} \text{ analog}}{V_{max} \text{ phe}} \times 100$
(C) $RCH_2CH(NH_2)CO_2H$	
R =	

	Compound[b]	$\dfrac{V_{max} \text{ analog}}{V_{max} \text{ phe}} \times 100$
(i)	β-Thien-2-ylalanine (2.5 mM)	68
(ii)	β-Pyridin-2-ylalanine (5 mM)	50
(iii)	β-Pyrazol-1-ylalanine (5 mM)	10
(iv)	β-Pyrazol-3-ylalanine (5 mM)	0
(v)	β-1,2,4-Thiazol-3-ylalanine (5 mM)	0

[a] After Anderson and Fowden (1970a).
[b] Figures in parentheses represent concentration of L-forms present in assay mixtures.

phenylalanine and the analog, although the ratio K_m(phe)/K_m (analog) remained nearly constant. Enzyme from *A. californica* exhibited the lowest affinity for the amino acid substrates; it also differed from the other four enzymes in regard to V_{max}(analog)/V_{max}(phe) ratios. At enzyme-saturating concentrations, 2-amino-4-methylhex-4-enoic acid was activated only 30 percent as efficiently as phenylalanine by *A. californica* enzyme, whereas the other four *Aesculus* enzymes activated the two amino acids at almost identical rates (Table 6). There is then a partial discrimination, as measured by ATP-^{32}PP$_i$ exchange rates, by *A. californica* enzyme against a product almost unique to this species, and this selectively contributes to the exclusion of 2-amino-4-methylhex-4-enoic acid from protein of *A. californica* seed. Presumably, further selective discrimination against the analog can occur at the stage of aminoacyl residue transfer to phe-specific-tRNA. Some incorporation of 2-amino-4-methylhex-4-enoic acid residues into protein of

TABLE 6

KINETIC DATA DERIVED FOR PHENYLALANYL-tRNA SYNTHETASES FROM *Aesculus* spp.
REPRESENTING THE FIVE SUBGENERIC SECTIONS[a]

	K_m ATP(M)	K_m phe(M)	K_m Amha[b] (M)	K_m Amha[b]/ K_m phe	V_{max} Amha $\dfrac{V_{max}\ \text{phe}}{} \times 100$
Calothyrsus					
A. *californica*	3.3×10^{-4}	1.05×10^{-4}	3.2×10^{-3}	30	30
Pavia					
A. *glabra*	3.6×10^{-4}	4.9×10^{-5}	1.78×10^{-3}	36	90
Aesculus					
A. *hippo-* *castanum*	2.8×10^{-4}	3.1×10^{-5}	1.18×10^{-3}	38	100
Parryaneae					
A. *parryi*	3.8×10^{-4}	1.9×10^{-5}	8.9×10^{-4}	47	100
Macrothyrsus					
A. *parviflora*	2.8×10^{-4}	1.6×10^{-5}	6.8×10^{-4}	43	100

[a] After Fowden *et al.* (1970).

[b] Amha: abbreviation for 2-amino-4-methylhex-4-enoic acid.

developing mung bean seedlings takes place when seeds are germinated in the presence of the analog; the data in Table 4 suggest that phenylalanyl-tRNA synthetase from *P. aureus* has a higher relative affinity for the analog than that shown by *A. californica* enzyme.

On the basis of morphological evidence, Hardin (1957, 1960) has concluded that *Aesculus* species assigned to the section Macrothyrsus are more closely related to those in section Calothyrsus, than in section Pavia. A different conclusion is reached if the distribution of unusual amino acids among species forming the genus *Aesculus* is examined (Fowden *et al.*, 1970) because sections Macrothyrsus and Pavia are seen to have many similarities, but both groups differ sharply in composition from members of section Calothyrsus. The kinetic data derived for the different phenylalanyl-*t*RNA synthetases (Table 6) tend to confirm the closer relationship of Macrothyrsus and Pavia.

The same 800-fold purified phenylalanyl-tRNA synthetase from *A. hippocastanum* has been used to study the affinity of 1-amino-2-phenyl-ethane-1-phosphonic acid (**26**), the phosphonate analog of phenylalanine. Although this phosphonate caused no stimulation of ATP-^{32}PP$_i$ exchange, i.e., it showed no substrate character, it nevertheless behaved as a strong competitive inhibitor of the activation of phenylalanine. The calculated inhibitor constant (K_i) was 1.7×10^{-5} M, a value almost identical with the K_m (phenylalanine) of 1.6×10^{-5} M determined for this particular preparation of enzyme. Phosphonate analogs of other amino acids, and especially that of tyrosine, had no inhibitory effect upon the activity of the phenylalanyl-tRNA synthetase. Indeed, in our limited experience, 1-amino-2-phenylethane-1-phosphonic acid was unique among phosphonate analogs in its behavior as an inhibitor of aminoacyl-tRNA synthetases; leucyl-, valyl-, and tyrosyl-tRNA synthetases from *A. hippocastanum* were totally insensitive to the phosphonate derivatives of their natural amino acid substrates (Anderson and Fowden, 1970b). Professor E. Neuzil (personal communication), however, found that the phosphonic acid derivative of tyrosine acts as an inhibitor of tyrosyl-tRNA synthetase from *E. coli.* The interaction of tetrazole derivatives of phenylalanine (27), and of leucine and isoleucine, with aminoacyl-tRNA synthetases from *P. aureus* has been the subject of a similar brief study (R. D. Norris and L. Fowden, unpublished results). Compound **27** was less effective than the corresponding phosphonate as an inhibitor of phenylalanyl-tRNA synthetase and, like the phosphonate, showed no substrate activity. Inhibition was not fully specific, for the tyrosyl synthetase was inhibited to about the same extent as the phenylalanyl enzyme. Tetrazole derivatives of leucine and isoleucine also produced limited inhibition of a range of individual aminoacyl-tRNA synthetases.

Some very preliminary results obtained by R. Watson and L. Fowden (unpublished experiments) suggest that 3-hydroxymethylphenylalanine can act as a substrate for the tyrosyl-tRNA synthetase from mung bean: the V_{max} value determined for **6** was approximately 110 percent of the similar value calculated for tyrosine itself. The same compound was not significantly activated by the phenylalanyl-tRNA synthetase from *P. aureus.* When tyr-tRNA synthetase was prepared from *C. tinctoria* seedlings, 3-hydroxymethylphenylalanine was activated at about 30 percent of the rate measured for phenylalanine; hence there was a partial discrimination by the enzyme against the plant's own product. Molecular models indicate that the hydroxyl proton in **6** can attain a spatial orientation in relation to the aromatic ring almost identical with that normally taken up by the hydroxyl proton of tyrosine—perhaps one essential feature for the binding of substrates at the active site of tyrosyl-tRNA synthetase is a requirement for hydrogen bond formation through such a proton. 3-Hydroxymethyl-4-hydroxyphenylalanine, like 3,4-dihydroxyphenylalanine, was a relatively poor substrate for the tyrosyl-tRNA synthetases from both mung bean seed and *C. tinctoria* seedlings.

Proline Analogs and the Active Site of Prolyl-tRNA Synthetase

The previous sections describe work in which naturally occurring amino acids were shown to exhibit analog behavior in relation to the transport of phenylalanine across cell membranes, in the regulation of metabolic pathways implicated in aromatic amino acid biosynthesis, and in their role as competitive substrates for phenylalanyl-tRNA synthetase. The experiments with aromatic amino acids thus extended to another group of compounds concepts earlier advocated (Fowden *et al.*, 1967) to explain the growth-inhibitory behavior of azetidine-2-carboxylic acid upon many organisms. One of the more important findings emerging from these earlier studies with imino acid analogs was that sharp differences in the substrate specificity of prolyl-tRNA synthetase existed between enzymes isolated from plant sources: enzyme isolated from a plant producing A2C abundantly, e.g., *Convallaria majalis*, discriminated absolutely against this substrate, but enzyme from a nonproducer species, e.g., *Phaseolus aureus*, accepted A2C as a substrate. More recently, we have extended our work to include studies with enzymes from many more plant species, and also with another naturally occurring proline analog, *cis*-3,4-methanoproline from *Aesculus parviflora*. The new experiments have attempted to explain species differences in substrate specificity in terms of the molecular conformation at the enzymes' active sites.

The kinetic parameters governing the activation of A2C and 3,4-methanoproline by purified preparations of prolyl-tRNA synthetase from various plant sources are given in Table 7 (which abstracts data from Norris and Fowden, 1972). The first three genera either contain appreciable amounts

TABLE 7

KINETIC PARAMETERS DETERMINED FOR PROLYL-tRNA SYNTHETASE PREPARATIONS FROM VARIOUS PLANTS[a]

Plant species	L-Proline $K_m(\times 10^4)$	L-Azetidine-2-carboxylic acid		*cis*-3,4-Methano-L-proline	
		$K_m(\times 10^3)$	V_{max}[b]	$K_m(\times 10^3)$	V_{max}[b]
Parkinsonia aculeata	4.35	∞	0–5	7.1	42
Delonix regia	1.82	∞	0–5	4.6	22
Convallaria majalis	4.5	∞	0–5	2.5	36
Beta vulgaris	4.5	2.2	73	n.d.[c]	<3
Hemerocallis fulva	6.25	5.3	75	n.d.	<3
Phaseolus aureus	1.37	1.43	55	n.d.	<2
Ranunculus bulbosa	2.9	2.0	66	∞	0

[a] All data derived from measurements of ATP-^{32}PP$_i$ exchange stimulated by proline or its analogs.

[b] V_{max} values expressed in terms of V_{max}(proline) = 100.

[c] n.d., no determination.

of A2C in their seeds (*Convallaria*) or produce large amounts during early germination (*Parkinsonia* or *Delonix*); *Beta* is a genus elaborating trace amounts of the imino acid (see above, Distribution and Concentration); there is no record of A2C occurring in *Hemerocallis, Phaseolus,* or *Ranunculus. cis*-3,4-Methanoproline, an analog molecule somewhat larger than the parent molecule of proline, is not known as a constituent of any of these genera. Table 7 confirms that species producing large amounts of A2C possess a type of prolyl-tRNA synthetase that discriminates against the imino acid; enzymes from other species, including *Beta vulgaris,* accept A2C readily as a substrate, at least in the first phase of the activation process associated with aminoacyladenylate formation. *cis*-3,4-Methano-proline behaves in a contrasting manner, being activated by A2C-producing species, but being almost inactive as a substrate when tested with enzyme from other plants. Such observations suggest that the geometry of the active site of the prolyl-tRNA synthetase from plants discriminating against A2C is such that the binding of molecules larger than proline is possible. By displaying this flexibility, the fit of analogs smaller than proline is presumably too loose to ensure the correct configurational alignment of the analog's carboxyl group with the α-PO_4^{-3} group of ATP. Conversely, prolyl-tRNA synthetases activating A2C presumably possess smaller active sites that ensure correct ligand bonding of molecules smaller than proline, but prevent larger analogs from attaining the required positioning of their carboxyl groups.

Further studies on the two types of prolyl-tRNA synthetase have been confined to preparations obtained from *Phaseolus aureus* and *Delonix regia,* used as representative contrasting species. *Delonix* enzyme is considerably more thermolabile than enzyme from *Phaseolus* (compare Fig. 2): substrates confer considerable stabilization against thermal denaturation upon the enzymes, and this is particularly noticeable for the *Delonix* enzyme. Imino acid analogs can mimic the normal substrate proline by also affording significant protection to the enzymes against heat inactivation.

Kinetic analysis of thermal inactivation may be used to estimate protection constants (π) for enzyme–substrate (or analog) complexes in the absence of an enzyme reaction. The protection constant has been equated with the dissociation constant of the enzyme–protector complex (see Chuang and Bell, 1972). Values of the protection constants extrapolated to 45°C were determined for prolyl-tRNA synthetase preparations and indicated that the smaller analog molecule A2C protected the *Delonix* enzyme less efficiently (relative to proline) than the *Phaseolus* enzyme (Norris and Fowden, 1973a). This observation supports the view that the *Delonix* enzyme possesses a larger active site. Possibly the positioning of

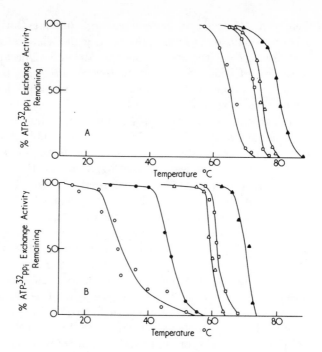

Fig. 2. Thermal inactivation profiles of prolyl-tRNA synthetase from *Phaseolus* (A) and *Delonix* (B) O———O, enzyme alone; ●———●, enzyme + 40 mM mercaptoethanol; △———△, enzyme + 4 mM ATP; □———□, enzyme + 50 mM proline; ▲———▲, enzyme + 4 mM ATP + 50 mM proline. All samples were subjected to 7 min heating periods at the various temperatures employed.

A2C on the *Delonix* enzyme results in an incorrect alignment of the carboxyl group with another important functional group within the enzyme, and so reduces the magnitude of a conformational change within the enzyme molecule that constitutes a necessary part of an efficient stabilization mechanism. In the presence of ATP, proline was bound more strongly to both the *Delonix* and *Phaseolus* enzymes than when ATP was absent. π Values for ATP were lower than those calculated for proline with both enzymes, and it seems probable that the nucleoside triphosphate binds to the synthetase before the imino acid substrate, thereby effecting a conformational change in the enzyme conducive to stronger binding of the imino acid substrates. The π(A2C)/π(Pro) ratio for the *Delonix* enzyme was considerably higher than the corresponding ratio calculated for the *Phaseolus* enzyme either in the presence or absence of ATP. Therefore the ability of the *Delonix* enzyme to discriminate against A2C may be seen as an inefficiency of the analog to elicit a suitable conformational change in the larger proline-binding site of this enzyme.

The large enthalpy changes occurring when proline, A2C, or ATP become bound to the prolyl-tRNA synthetases presumably indicate that binding is associated with a high degree of ordering of the intramolecular protein structure, and therefore support the postulate that imino acid binding elicits a conformational change within the active site. A comparison of the enthalpy changes associated with binding of proline and A2C in the presence of ATP indicates that the analog causes relatively less intramolecular ordering of the *Delonix* enzyme than of the *Phaseolus* enzyme, but it has not been proved conclusively that such differences fully explain the discriminatory behavior of prolyl-tRNA synthetase from *Delonix* (see Norris and Fowden, 1973a).

In a related study, Norris and Fowden (1973b) have provided evidence indicating that a histidine residue and a cysteinyl –SH group are present within the binding site of prolyl-tRNA synthetase from both *Phaseolus* and *Delonix*. Both enzymes undergo methylene blue-mediated photoinactivation. Light alone caused some inactivation of the *Delonix* enzyme. Proline, A2C, and several other proline analogs afforded considerable protection of the enzyme against dye-mediated photoinactivation, but ATP was ineffective. This observation was consistent with the view that a histidinyl residue within the active site plays an important role in the binding of the imino acid substrate. Further support for this concept was gained from an experiment to study the variation of K_m (proline) with pH determined for the ATP-^{32}PP$_i$ exchange catalyzed by prolyl-tRNA synthetase. Figure 3

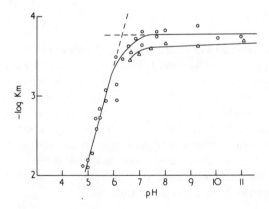

FIG. 3. Effect of pH on the K_m values determined for proline in the ATP-^{32}PP$_i$ exchange reaction catalyzed by the prolyl-tRNA synthetase from *Phaseolus* (O——O) and *Delonix* (△——△).

shows the relationships obtained for enzymes from *Phaseolus* and *Delonix*. Inflections are seen clearly at pH values slightly above 6.0. It is most unlikely that either the imino acid substrate (pK_a values for proline are 1.92 and 9.73) or ATP (as an ATP-Mg-PP$_i$ complex, the pK_a is less than 4.0) would undergo an ionization change near pH 6, and the group most likely to show a sharp ionization change at this pH would be the imidazole moiety (pK_a 6.09) of a histidine residue within the enzyme itself.

Prolyl-tRNA synthetase preparations from both species were strongly inhibited by *p*-chloromercuribenzoate (*p*CMB). The inhibited enzymes could be reactivated by addition of sulphydryl reducing reagents, while the degree of inhibition caused by *p*CMB could be markedly reduced if the enzymes were protected by addition of proline, A2C, ATP, or *t*RNA. All these facts are consistent with the idea that a cysteinyl-SH group is present within the active site of the enzymes.

The mechanistic role of the histidine and cysteine residues at the active site remains in doubt. Perhaps nucleophilic attack by the imidazole N at the carbonyl oxygen of proline may be responsible for the formation of an acyl bond with the α-PO$_4^{3-}$ group of ATP. Additional nucleophilic catalysis may be provided by the SH group. The ability of the *Delonix* enzyme to discriminate against A2C may be related to the positions of the histidine and cysteine groups relative to the imino acid binding site. Possibly, the alignment of the analog's carboxyl group relative to the SH or imidazole groups may minimize the nucleophilic catalytic action of these groups on A2C, while the alignment of the carboxyl of proline meets the requirements for efficient promotion of ATP-^{32}PP$_i$ exchange.

Summary

During the past 25 years, the number of amino acids known to occur in plants has dramatically increased. While many of the compounds remain as metabolic curiosities, having no established physiological role, others have been used as analogs to shed important information upon the uptake of amino acids across cell membranes, upon the regulatory control mechanisms governing amino acid biosynthesis, and upon the specificity of enzymes involved in the early stages of protein synthesis. Interest in the nonprotein amino acids then has broadened, and they now receive attention not just from plant biochemists interested in secondary product metabolism, but also from geneticists and physiological ecologists.

Acknowledgment

A grant from NATO facilitated some aspects of the work described in this review.

REFERENCES

Adinolfi, M. 1964. *Rend. Accad. Sci. Fis. Mat., Naples* [4] **31**:335–337.
Ames, G. F. 1964. *Arch. Biochem. Biophys.* **104**:1–18.
Ames, G. F., and J. R. Roth. 1968. *J. Bacteriol.* **96**:1742–1749.
Anderson, J. W., and L. Fowden. 1970a. *Biochem. J.* **119**:677–690.
Anderson, J. W., and L. Fowden. 1970b. *Chem.-Biol. Interact.* **2**:53–55.
Bell, E. A. 1962. *Biochem. J.* **83**:225–229.
Bell, E. A. 1972. *In* "Phytochemical Ecology" (J. B. Harborne, ed.), pp. 163–177. Academic Press, New York.
Bell, E. A., and L. Fowden. 1964. *In* "Taxonomic Biochemistry and Serology" (C. A. Leone, ed.), pp. 203–223. Ronald Press, New York.
Bell, E. A., and D. H. Janzen. 1971. *Nature (London)* **229**:136–137.
Bell, E. A., and A. S. L. Tirimanna. 1965. *Biochem. J.* **97**:104–111.
Brown, K. D. 1970. *J. Bacteriol.* **104**:177–188.
Chilton, W. S., G. Tsou, L. Kirk, and R. G. Benedict. 1968. *Tetrahedron Lett.* pp. 6283–6284.
Chuang, H. Y. K., and F. E. Bell. 1972. *Arch. Biochem. Biophys.* **152**:502–514.
Conway, T. W., E. M. Lansford, and W. Shive. 1962. *J. Biol. Chem.* **237**:2850–2854.
Dardenne, G. A., J. Casimir, and J. Jadot. 1968. *Phytochemistry* **7**:1401–1406.
Dardenne, G. A., M. Marlier, and J. Casimir. 1972. *Phytochemistry* **11**:2567–2570.
Daxenbichler, M. E., C. H. VanEtten, E. A. Hallinan, and F. R. Earle. 1971. *J. Med. Chem.* **14**:463–465.
Dietrichs, H. H., and H. Funke. 1967. *Holzforschung* **21**:102–107.
Done, J., and L. Fowden. 1952. *Biochem. J.* **51**:451–458.
Dunnill, P. M., and L. Fowden. 1965. *Phytochemistry* **4**:933–944.
Dunnill, P. M., and L. Fowden. 1967. *Phytochemistry* **6**:1659–1663.
Eagles, J., W. M. Laird, S. Matal, R. Self, and R. L. M. Synge. 1971. *Biochem. J.* **121**: 425–430.
Fangman, W. L., and F. C. Neidhardt. 1964. *J. Biol. Chem.* **239**:1839–1843.
Floss, H. G., D. K. Onderka, and M. Carroll. 1972. *J. Biol. Chem.* **247**:736–744.
Fowden, L. 1962. *Endeavour* **81**:35–42.
Fowden, L. 1963. *J. Exp. Bot.* **14**:387–398.
Fowden, L. 1965. *Sci. Progr. (London)* **53**:583–599.
Fowden, L. 1970. *Progr. Phytochem.* **2**:203–266.
Fowden, L. 1972a. *Phytochemistry* **11**:2271–2276.
Fowden, L. 1972b. *Carbon-Fluorine Compounds, Ciba Found. Symp.* pp. 141–159.
Fowden, L., and H. M. Pratt. 1973. *Phytochemistry* **12**:1677–1681.
Fowden, L., and A. Smith. 1968. *Phytochemistry* **7**:809–819.
Fowden, L., and F. C. Steward. 1957. *Ann. Bot. (London)* [N.S.] **21**:53–67.
Fowden, L., D. Lewis, and H. Tristram. 1967. *Advan. Enzymol.* **29**:89–163.
Fowden, L., J. W. Anderson, and A. Smith. 1970. *Phytochemistry* **9**:2349–2357.
Guroff, G., J. W. Daly, D. M. Benson, B. Witkop, and S. Udenfriend. 1967. *Science* **157**: 1524–1530.
Hadwiger, L. A., H. G. Floss, J. R. Stoker, and E. E. Conn. 1965. *Phytochemistry* **4**:825–830.
Hardin, J. W. 1957. *Brittonia* **9**:173–195.
Hardin, J. W. 1960. *Brittonia* **12**:26–39.
Hill, R. K., and G. R. Newkome. 1969. *J. Amer. Chem. Soc.* **91**:5893–5894.
King, F. E., T. J. King, and A. J. Warwick. 1950. *J. Chem. Soc., London* pp. 3590–3597.
Kjaer, A., and P. O. Larsen. 1963. *Acta Chem. Scand.* **17**:2397–2409.
Larsen, P. O. 1967. *Biochim. Biophys. Acta* **141**:27–46.
Larsen, P. O., and H. Sørensen. 1968. *Biochim. Biophys. Acta* **156**:190–191.

Larsen, P. O., D. K. Onderka, and H. G. Floss. 1972. *J. Chem. Soc., Chem. Commun.* pp. 842–843.

Liss, I. 1961. *Flora (Jena)* 151:351–367.

Morris, C. J., and J. F. Thompson. 1965. *Arch. Biochem. Biophys.* 110:506–510.

Morris, C. J., J. F. Thompson, S. Asen, and F. Irreverre. 1959. *J. Amer. Chem. Soc.* 81:6069–6070.

Mothes, K., H. R. Schütte, P. Müller, M. von Ardenne, and R. Tümmler. 1964. *Z. Naturforsch.* B19:1161–1162.

Müller, P., and H. R. Schütte. 1967. *Flora (Jena)* 158:421–432.

Nair, P. M., and L. C. Vining. 1965. *Phytochemistry* 4:401–411.

Norris, R. D., and L. Fowden. 1972. *Phytochemistry* 11:2921–2935.

Norris, R. D., and L. Fowden. 1973a. *Phytochemistry* 12:2109–2122.

Norris, R. D., and L. Fowden. 1973b. *Phytochemistry* 12:2829–2841.

Onderka, D. K., and H. G. Floss. 1969. *J. Amer. Chem. Soc.* 91:5894–5896.

Schneider, G. 1958. *Biochem. Z.* 330:428–432.

Seneviratne, A. S., and L. Fowden. 1968. *Phytochemistry* 7:1039–1045.

Sloane, N. H., and S. C. Smith. 1968. *Biochim. Biophys. Acta* 158:394–401.

Smith, I. K., and L. Fowden. 1968. *Phytochemistry* 7:1065–1075.

Smith, L. C., J. M. Ravel, S. R. Lax, and W. Shive. 1962. *J. Biol. Chem.* 237:3566–3570.

Sung, M. L., and L. Fowden. 1969. *Phytochemistry* 8:2095–2096.

Sung, M. L., L. Fowden, D. S. Millington, and R. C. Sheppard. 1969. *Phytochemistry* 8:1227–1233.

Thompson, J. F., C. F. Morris, S. Asen, and F. Irreverre. 1961. *J. Biol. Chem.* 236:1183–1185.

Torquati, T. 1913. *Arch. Farmacol. Sper. Sci. Affini.* 15:308–312.

Tristram, H. 1973. *In* "Biosynthesis and its Control in Plants" (B. V. Milborrow, ed.), pp. 21–48. Academic Press, New York.

Turner, B. L., and J. B. Harborne. 1967. *Phytochemistry* 6:863–866.

Watson, R., and L. Fowden. 1973. *Phytochemistry* 12:617–622.

Willshaw, G., and H. Tristram. 1972. *Biochem. J.* 127:71 p.

PROTEINASE INHIBITORS IN NATURAL PLANT PROTECTION*

C. A. RYAN and T. R. GREEN†

Department of Agricultural Chemistry,
Washington State University, Pullman, Washington

Introduction

Naturally occurring proteinase inhibitors are polypeptides and proteins with molecular weights ranging from 4000 to over 50,000 that are apparently ubiquitous in nature (Vogel et al., 1969; Laskowski and Sealock, 1971). They usually interact with, and inhibit, the animal pancreatic proteinases chymotrypsin and trypsin, or enzymes with similar specificities. Their interactions with proteinases are very specific and they bind very tightly to the proteinases they inhibit, having K_I's of 10^{-8} to 10^{-11} M.

* Scientific paper 4122, Project 1791. College of Agriculture Research Center, Washington State University, Pullman, Washington.
† Present address: Department of Chemistry, University of California at Los Angeles, Los Angeles, California.

In animals they have been implicated in regulating endogenous proteolytic enzymes and in protecting tissue or fluid proteins from unwanted or foreign proteolysis (Werle and Zickgraf-Rudel, 1972).

In plants their function has not been clearly established. The proteinase inhibitors present in plants have a specificity almost exclusively for serine proteinases that are found in animals and microorganisms but that are only rarely found in plants (Ryan, 1973). This is particularly interesting because some seeds and tubers contain very large amounts of the inhibitors. For example, the water-soluble proteins of soybeans contain over 6 percent trypsin inhibitors (Rackis and Anderson, 1964), and the soluble proteins of barley grains (Mikola and Kirsi, 1972) and of the apical cortex of potatoes contain over 10 percent proteinase inhibitors (C. A. Ryan, unpublished data).

The roles of proteinase inhibitors in plants have been thought to be both regulatory and protective, and several workers have suggested both of these possibilities in the past. Proteolytic enzymes in several germinating seeds have been shown to be inhibited by proteinase inhibitors during germination (Shain and Mayer, 1965; Seidel and Jaffe, 1967; Polanowski, 1967; Burger and Siegelman, 1966; Mikola and Enari, 1970) and these enzymes are probably regulated by the inhibitor. However, inhibitors of plant proteinases represent only a very small percentage of total inhibitors and a regulatory role for the bulk of inhibitors does not seem likely.

The possible role of inhibitors as protective agents against insects has been strongly implied (Mickel and Standish, 1947; Lipke *et al.*, 1954; Birk *et al.*, 1963; Applebaum *et al.*, 1964). The work of Birk and Applebaum demonstrated that specific inhibitors of a number of insect larval gut proteinases are present in seeds, and that several well-known pure plant proteinase inhibitors are effective inhibitors of insect proteinases. Applebaum (1964) argued that proteinase inhibitors were protective agents and proposed that legumes evolved proteinase inhibitors as a defense mechanism against insects and that the inhibition of digestion should be considered as an important factor in host selection.

Proteinase inhibitors in foodstuffs may have had important effects toward animals other than insects. It has been noted (Leopold and Ardrey, 1972) that proteinase inhibitors are the most common toxic substances that occur in plants and they may have had considerable effect in restricting food, even for primitive man.

Our interest in proteinase inhibitors began with the discovery and isolation of Chymotrypsin Inhibitor I from potato tubers by Balls and Ryan in 1962 (Ryan and Balls, 1962; Balls and Ryan, 1963). This has subsequently been studied extensively by Ryan and his associates (Ryan,

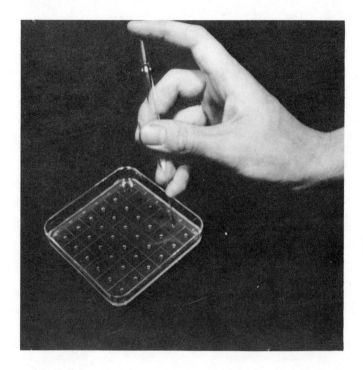

FIG. 1. Immunological radial diffusion agar plate for assaying Inhibitor I. The agar contains anti-Inhibitor I serum. Wells are filled with solutions containing Inhibitor I. As the inhibitor protein migrates radially into the agar, precipitin rings result whose diameters are a function of the concentration of Inhibitor I in the solutions introduced into the well.

1966; Melville and Ryan, 1972). Some of the more important properties of this protein are shown in Table 1.

In order to study this one inhibitor among all of the other plant proteins we have employed a specific quantitative immunological technique (Ryan, 1967) that utilizes specific rabbit antibodies raised against pure Inhibitor I. The anti-Inhibitor I serum is incorporated into melted, cooling agar, and poured into petri dishes. When the agar has solidified, holes are made in it using a machined punch, followed by removal of the agar from the wells with suction. The wells are exactly filled with standards or the extracts to be tested. Figure 1 shows a typical agar plate having its wells filled with a sample of tomato leaf juice. As the antigen (Inhibitor I) in the wells migrates radially into the agar, precipitin rings develop whose diameters when squared are a linear function of the concentration of Inhibitor I present in the sample, as shown in Fig. 2. Thus, the assay

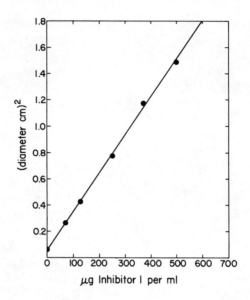

Fig. 2. A plot of the diameters squared vs. concentration of Inhibitor I in solutions in the radial diffusion assay.

TABLE 1

Potato Inhibitor I

Molecular weight	
Tetramer	39,000
Monomer	9,800
Complex with chymotrypsin	140,000
Inhibits	
Chymotrypsin	$K_I = 1.3 \times 10^{-11}\ M$
Trypsin	—
Subtilisin	—
Pronase (one species)	—
Does not inhibit sulfhydryl, metallo, or acid proteinases	
Stability	
Stable at pH 2 to 10 at 80°C, 10 minutes	
Resistant to proteolysis at neutral pH by every endopeptidase or exopeptidase tested to date	
Origin	
Cortex and pith of Russet potatoes	
Can comprise up to 7 percent of soluble proteins of apical cortex	

allows the absolute concentrations of Inhibitor I to be determined directly in virtually any solution of near-neutral pH, including tissue juice or crude extracts.

We began our study by assaying Inhibitor I in tissues of potato plants to see if any relationship could be found between the concentration of inhibitor and physiological events of growth and development. We found that Inhibitor I was a transient component of potato tissues and it usually appeared in leaves just prior to the breaking of apical dominance, or, as in the case of young potato plants, the establishment of rhizomes. In potato tubers it was found to be concentrated in the apical cortex and it disappeared during sprouting, accumulating in young sprouts and in young expanding leaves. It subsequently disappeared from leaves when tuberization began, accumulating in the new tubers. A review of this research can be found elsewhere (Ryan and Shumway, 1971).

Discovery of the Wound Response

During the study it was noted that tomato leaves also possessed Inhibitor I, but its presence was unpredictable. We often found that leaves of some tomato plants exhibited extraordinarily high concentrations of Inhibitor I whereas leaves of adjacent plants had little or no inhibitor. It

TABLE 2

Colorado Potato Beetle–Induced Accumulation of
Chymotrypsin Inhibitor I in Tomato Plants

	Average Inhibitor I concentration (μg/ml) in		
	Leaves	Main stem	Roots
Beetle damaged leaves[a]	202 (77–235)	52 (0–73)	<15
Controls, no damage	47 (0–120)	<15	<15

[a] Each value is an average from 11 trifoliate leaflets from the second leaf down from apex. Adult beetles were allowed to feed randomly on plants for 24 hours. After an additional 24 hours the tissues were assayed for Inhibitor I immunologically. Experiments were carried out in a greenhouse under natural light. Leaf damage varied from minor damage to a single leaflet to severe damage to all leaflets. The accumulation of inhibitor in leaves varied, in general, proportionally to the insect damage inflicted on the plants (Green and Ryan, 1972). Copyright 1972 by the American Association for the Advancement of Science.

FIG. 3. Colorado potato beetle.

was decided that perhaps the reason might be that the plants were in-fected with a pathogen or that they might have been attacked by insects. This hypothesis was tested (Green and Ryan, 1972) by allowing insects to feed on potato or tomato plants. This would cause wounding and at the same time the wound would be inoculated with bacteria and fungi by the insect mouth parts. We allowed Colorado potato beetles (Fig. 3), a common insect on Solanaceous plants in the Pullman area, to feed on leaves of several young tomato plants for 24 hours. The plants were then rested for 24 hours. The leaves of the beetle-infested plants contained considerably more inhibitor than uninfested plants (Table 2). Similar results were obtained with young potato plants. Assays of individual leaves showed that undamaged leaves of infested plants had accumulated significant Inhibitor I as well as the damaged leaves.

We found that damage to a single leaf was enough to cause all of the leaves of small plants to accumulate Inhibitor I. This suggested that a signal was transmitted from the wound site throughout the plants which caused the accumulation of Inhibitor I.

Characterization of the Wound Response

The response invoked by insect wounding could be simulated by mechan-ically wounding leaves in a variety of ways. A crushing or grinding action of any kind was effective. Reasonably reproducible results were obtained with a paper punch or by crushing the leaf between the flat end of a small wooden rod and a flat file. Figure 4 shows the relationship between the number of wounds administered in these ways and the amount of Inhibitor I that was induced to accumulate in a leaf adjacent to the wounded leaflet, 48 hours after wounding (Green and Ryan, 1972).

The location of the wound on the leaf surface was important in determining the accumulation of Inhibitor I in leaves adjacent to those wounded. Figure 5 shows a relationship of large and small wounds and their location with the resultant Inhibitor I accumulation. A large wound on the main vein was most effective, indicating that the release of the proteinase inhibitor inducing factor (PIIF) near a main vein facilitated its transport out of the leaf. Transport of PIIF out of the leaf into the rest of the plant could be prevented by excising the wounded leaf with a clean slice of the petiole with a razor blade. This slice did not release enough PIIF into the plant to induce accumulation of Inhibitor I. With this technique the rate of transport of PIIF from the wound site through the petiole could be determined.

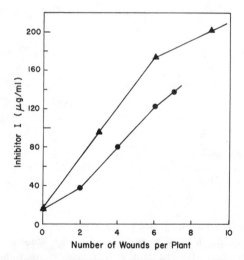

Fig. 4. The accumulation of Inhibitor I in leaf juice of the tomato plant as a result of mechanical wounding by crushing leaves between a wooden dowel and file ▲——▲; and by punching holes in leaflets with a paper punch ●——●. The tomato plants were young, 8 to 10 cm in height, with two well-developed leaves, a lower trifoliate, and an upper adjacent pentafoliate leaf. The lower leaf was wounded to initiate the experiment. Plants were maintained under midsummer greenhouse conditions for 48 hours. The terminal leaflet of the pentafoliate leaf was assayed immunologically for Inhibitor I (Green and Ryan, 1972). Copyright 1972 by the American Association for the Advancement of Science.

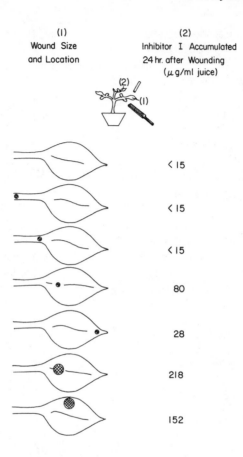

Fɪɢ. 5. The relationship of the location and relative size of wounds to the quantity of Inhibitor I induced to accumulate in adjacent leaves 48 hours after wounding (C. A. Ryan, unpublished results).

Just after wounding single leaves, and at intervals for several hours thereafter, wounded leaves were excised. The amount of Inhibitor I accumulated in the adjacent upper leaf was measured 48 hours after wounding. Figure 6 shows that excision of wounded leaves immediately after wounding prevented PIIF from leaving the leaves. At increasingly longer times after wounding, excision was decreasingly effective in preventing PIIF from being exported (Green and Ryan, 1973). Figure 6 also shows that the half-time for export of PIIF out of wounded leaves was about 1 hour. This represents a velocity of travel of PIIF of about 2–3 cm/hour.

The time course of the overall wound response was determined by

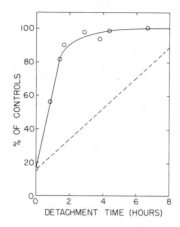

Fig. 6. The 48-hour accumulation of Inhibitor I in young tomato terminal leaflets from leaves adjacent to leaves wounded at zero time. The wounded leaves were detached from the plant at the times shown in an effort to prevent transport of PIIF. Details are given in the text; O——O, 1000 ft-c, 30°C; – – –, greenhouse conditions (Green and Ryan, 1972). Copyright 1972 by the American Association for the Advancement of Science. Inhibitor I concentration was determined immunologically (Green and Ryan, 1973).

measuring Inhibitor I in leaves adjacent to leaves wounded at time zero. As shown in Fig. 7, within 12 hours Inhibitor I usually could be detected and the accumulation often continued for over 100 hours (Green and Ryan, 1972). The accumulation usually ceased 3 or 4 days after severe

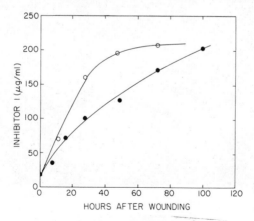

Fig. 7. Time course of accumulation of Inhibitor I in young tomato terminal leaflets from leaves adjacent to leaves wounded at zero time; O——O, 1000 ft-c, 30°C; ●——●, greenhouse conditions, late August. The hatched area represents the variability found in control plants maintained in constant darkness. Inhibitor I concentration was determined immunologically (Green and Ryan, 1973).

wounding, or earlier if the wound was not severe and was not reinforced. However, the accumulated inhibitor remained in the leaves for weeks, its level decreasing slowly with time.

The wound response is entirely dependent upon light and is temperature-dependent. The maximum accumulation takes place above 400 foot-candles at a temperature of 36°C.

Proteinase Inhibitor Inducing Factor (PIIF)

Assay System with Excised Tomato Leaves

Following up our earlier observation that single leaves could be excised cleanly with a sharp razor blade without releasing significant quantities of PIIF into the plant, we found that excised leaves could be supplied with water and incubated in light without accumulating Inhibitor I. When these leaves were supplied briefly (10–15 minutes) with crude tomato leaf juice before placing in water and light, the leaves accumulated large quantities of Inhibitor I within 24 hours. The system is illustrated in Fig. 8. Thus, leaves from very young tomato plants provide a simple,

Fig. 8. The assay of PIIF with leaves of young tomato plants. Leaves were excised and supplied with PIIF-containing solutions (left) for 10–15 minutes, then transferred to water (right) and incubated in light at 800 ft-c and 31°C for 24 hours. The leaves were then assayed immunologically for Inhibitor I content (C. A. Ryan, unpublished data).

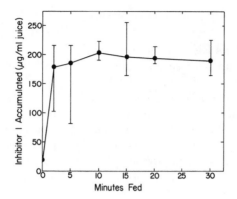

Fig. 9. Accumulation of Inhibitor I in detached tomato leaves as a function of time of uptake of crude tomato leaf juice. After imbibing juice for the times indicated the leaves were transferred to water for 24 hours at 800 ft-c and 31°C and assayed for Inhibitor I. Vertical lines indicate ranges of data (C. A. Ryan, unpublished data).

direct, controlled assay system for studying the PIIF-induced accumulation of Inhibitor I.

It takes but 10–15 minutes to saturate a leaf with the PIIF contained in crude tomato juice (Fig. 9). It was calculated, using tritiated water, that a small leaf takes up about 0.5 to 1 μl of solution per minute. Thus, less than 10 μl of crude, centrifuged tomato juice contains enough PIIF to cause an entire leaf to accumulate over 200 μg of Inhibitor I in 24 hours.

A number of properties of the PIIF-induced accumulation of Inhibitor I in excised tomato leaves was studied. The time course of the reaction (Fig. 10) indicated that the accumulation of Inhibitor I did not commence

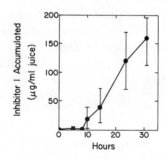

Fig. 10. Time course accumulation of Inhibitor I in detached leaves supplied with clarified juice from steamed tomato leaves. Juice was imbibed for 15 minutes followed by water for under 800 ft-c at 31°C and assayed for Inhibitor I at the times indicated. Vertical lines indicate ranges of data (C. A. Ryan, unpublished data).

immediately after taking up the PIIF-containing juice. A lag of 8–10 hours ensued that was followed by a linear increase in Inhibitor I for several hours. This lag is reminiscent of the lag that occurs when barley aleurone layers are treated with gibberellic acid (Jacobsen and Varner, 1966). During the early stages of gibberellin-induced amylase synthesis there appear dramatic changes of a number of cellular components including ribosomes that precede amylase synthesis (Evins and Varner, 1972). The leaf system studied here, however, is quite different in that the metabolism of growing tissue cells is being shifted drastically into producing large quantities of a single protein, whereas the barley seed is a dormant organ. On the other hand, these apparent differences may have similarities at the cellular level.

It was of primary interest to know if any of the known plant hormones, including gibberellin, were responsible for the accumulation of Inhibitor I in the excised leaf assay system. Table 3 shows that none of the six plant hormones tested with young tomato leaves could simulate PIIF or significantly antagonize its action. It was of interest that gibberellic acid as well as traumatin, the wound hormone that is released from wounded tissues of certain plants, were completely inactive in this system.

Protein synthesis inhibitors were also supplied to leaves for 20 minutes before supplying PIIF for 15 minutes. Table 4 demonstrates that actinomycin D and cycloheximide were effective inhibitors of the PIIF-induced accumulation of Inhibitor I, but that rifampin and chloramphenicol had no effect. This indicated that RNA synthesis and cytoplasmic ribosomes were somehow involved in the accumulation and that inhibition of chloroplast or mitochondrial protein synthesis did not prevent the response.

ISOLATION OF PIIF

From experiments performed previously (Ryan and Huisman, 1970) we knew that injuring leaf petioles with live steam caused a large increase in Inhibitor I in intact plants within 48 hours after steaming. The steaming was evidently as effective as mechanical wounding in releasing PIIF from the tissues. We found that placing whole tomato plants in live steam resulted in excellent recoveries of PIIF from the juice of the steamed plants. Autoclaving the plants for 10–15 minutes gave the same results. Thus, a crude stock of PIIF was prepared by autoclaving tomato leaves and freeze-drying them. The dried leaves were homogenized in water and centrifuged at 15,000g. The solution was a very active source of PIIF. As shown in Fig. 11, from 30–50 mg of dry leaf per milliliter gave a maximum induction of Inhibitor I accumulation when supplied to young

TABLE 3

EFFECT OF VARIOUS PLANT HORMONES ON PIIF-
INDUCED ACCUMULATION OF INHIBIROR I IN
EXCISED TOMATO LEAVES

Hormone (10^{-4} M)[a]	Inhibitor I Accumulated per 24 hour (μg/ml juice)	
	Hormone + H$_2$O	Hormone + PIIF (15 min)
None	29	242
Abcissic acid	<15	298
Ethylene	<15	283
Gibberellic acid (GA$_3$	<15	167
Indoleacetic acid	15	276
Kinetin	25	209
Traumatin	<15	255

[a] Hormone fed 10 minutes followed by crude
PIIF for 15 minutes. Incubated at 1000 ft-c at 30°C
(C. A. Ryan, unpublished data).

TABLE 4

EFFECT OF PROTEIN SYNTHESIS
INHIBITORS ON PIIF-INDUCED
INHIBITOR I ACCUMULATION IN
EXCISED TOMATO LEAVES

Additions[a] (10^{-4} M)	Inhibitor I Accumulation (μg/ml juice)
None	180
Actinomycin D	48
Cyoheximide	0
Rifampin	189
Chloramphenicol	209

[a] Inhibitor fed 20 minutes followed by crude PIIF for 15
minutes. Incubated at 1000 ft-c
at 30°C (C. A. Ryan, unpublished data).

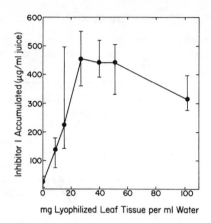

FIG. 11. Accumulation of Inhibitor I in detached tomato leaves as a function of lyophilized tomato leaf tissue solubles. Varying quantities of lyophilized leaf tissue were dispersed in water and centrifuged at 15,000*g*. The clarified leaf juice was supplied to detached young tomato plants through the cut petiole for 15 minutes and transferred to water. Conditions of incubation and assay are the same as in Fig. 9 (C. A. Ryan, unpublished results).

tomato leaves for 15 minutes, followed by incubation of the leaves in 800 foot-candle light at 31°C.

With the crude stock of autoclaved, freeze-dried tomato leaves as a starting material, we began attempts to purify PIIF using excised young tomato leaves as an assay system. PIIF was found to be insoluble in lipid solvents and very soluble in water. Therefore, the dried leaves were extracted with chloroform–methanol (2:1, v/v) to remove lipids, dried, and extracted with water. The water extract was reduced in volume and desalted with Biogel P2. PIIF eluted as a broad peak very near the void. The desalted crude PIIF was passed through QAE-Sephadex in dilute Tris buffer at pH 8.1. PIIF was not adsorbed and it eluted with the void volume. This step removed most of the colored materials. The entire void was desalted again through Biogel P2. It eluted as a double peak, one at the void and a later broad peak. The breakthrough peak contained most of the PIIF activity and was lyophilized. This material was highly active and represents our purest preparation of PIIF to date. This preparation of PIIF contains no detectable phosphate- or ninhydrin-reactive materials. It does, however, contain U.V.-absorbing material with a maximum near 280 nm and a large percentage of carbohydrate. The solution behavior on Biogel P2 suggests that it has a molecular weight above 2000. Its further purification and its composition are under study.

TABLE 5

The Identification of PIIF in Leaves of Various Higher Plants and in Fungal Fruiting Bodies[a]

Source of PIIF	PIIF activity (percent of the activity of tomato PIIF with excised tomato leaves)
HIGHER PLANTS	
Solanaceae	
Tomato	100
Potato	110
Tobacco	122
Nightshade	83
Datura	127
Petunia	27
Gramineae	
Corn	96
Barley	113
Wheat	102
Leguminosae	
Alfalfa	81
Pea	80
Broad bean	41
Honey locust	30
Cruciferae	
Radish	30
Rosaceae	
Strawberry	48
Apple	93
Anacardiaceae	
Sumac	7
Asteraceae (compositae)	
Lettuce	100
Ericaceae	
Indian pipe	80
Pinaceae	
White Pine	127
Yellow Pine	11
Polypodiaceae	
Fern	125
Equisetaceae	
Horsetail	105
FUNGI	
Agaricaceae	
Agaricus campestris	140
Boletaceae	
Boletus felleus	113

[a] D. McFarland and C. A. Ryan, unpublished data.

Phylogenetic Search for PIIF

We have demonstrated that PIIF is a component of the leaves of a number of plants in addition to tomato and potato. A survey of over 20 plant species (C. A. Ryan and D. McFarland, unpublished data) has shown that the leaf juice of all of the genera tested except one, the sumac *Rhus* of the Anacardiaceae, contains a factor(s) that causes tomato leaves to accumulate Inhibitor I (Table 5). Leaves from each species were subjected to live steam and then the juice was recovered and clarified at 15,000g. The clear juices were supplied to young tomato leaves through the cut petioles for 15 minutes as described in the previous sections. It is of particular note that the fern, horsetail, and fungi studied are phylogenetically a long distance from tomato but contained significant quantities of PIIF. This suggests that PIIF is widespread in nature. From these data we conclude that it is possible that the wound response may very well be a common property of plants.

Summary

The discovery that insect wounding or mechanical wounding of tomato or potato leaves results in the release of a proteinase inhibitor inducing factor (called PIIF) that is rapidly translocated throughout the plants, where it induces rapid accumulation of substance potentially harmful to insects and microorganisms, has many implications and ramifications.

The response appears to be a primitive immune-type response. It is possible that the plant senses insect attack through the release of PIIF and that proteinase inhibitors are produced to arrest proteolysis, not only by the insects but by microorganisms as well. It seems that such a response would not be effective against random attacks by flying insects unless the attacks were frequent. Defense against larval attack seems more likely. Larvae hatching from eggs deposited on the leaves would have little recourse but to eat the leaves. If the inhibitors were present then either death or serious impairment of normal development could ensue.

We have as yet no knowledge of specific insects being repelled by the presence of proteinase inhibitors as a result of wounding but are actively seeking such a relationship. We have initiated a genetic study to seek mutants with exceptional capacity to accumulate Inhibitor I, as well as those that only weakly accumulate Inhibitor I upon wounding. If we can locate such mutants they can be field-tested for changes in their resistance or susceptibility to insect predators. If we can demonstrate that such a system is operating effectively in nature, then its possible application in improving methods of biological pest control should be fully explored. Apart from its possible application toward biological pest control the wound response in tomato leaves appears to be a simple, uncomplicated system for studying the regulation of protein synthesis and degradation in plants. There are several immediate problems to consider: the cellular origins and location of PIIF, or its precursors; the mechanism of release and transport; and the mechanism of the regulation of Inhibitor I accumulation by PIIF. The isolation and synthesis of PIIF should be instrumental in approaching all of these problems.

ACKNOWLEDGMENTS

Financial assistance from the National Institute of Health, U. S. Public Health Service Grant 2-K3-GM 17059; National Science Foundation Grant GB 37972; and U. S. Department of Agriculture, Cooperative States Research Service Grants 915-15-79 and 316-15-60 is gratefully acknowledged. We thank Mr. Charles Oldenberg for growing our plants.

REFERENCES

Applebaum, S. W. 1964. *J. Insect Physiol.* **10**:783.
Applebaum, S. W., Y. Birk, I. Harpaz, and A. Bondi. 1964. *Comp. Biochem. Physiol.* **11**:85.
Balls, A. K., and C. A. Ryan. 1963. *J. Biol. Chem.* **238**:2976.
Birk, Y., A. Gertler, and S. Khalef. 1963. *Biochim. Biophys. Acta* **67**:326.
Burger, W. C., and H. W. Siegelman. 1966. *Physiol. Plant.* **19**:1089.
Evins, W. H., and J. E. Varner. 1972. *Plant Physiol.* **49**:348.
Green, T. R., and C. A. Ryan. 1972. *Science* **175**:776.
Green, T. R., and C. A. Ryan. 1973. *Plant Physiol.* **51**:19.
Jacobsen, J. V., and J. E. Varner. 1966. *Plant Physiol.* **42**:1596.
Laskowski, M., Jr., and R. W. Sealock. 1971. *In* "The Enzymes" (P. Boyer, ed.), 3rd ed., Vol. 3, pp. 375–473. Academic Press, New York.
Leopold, A. C., and R. Ardrey. 1972. *Science* **176**:512.
Lipke, H., G. S. Fraenkel, and I. E. Leiner. 1954. *J. Agr. Food Chem.* **2**:410.
Melville, J. C., and C. A. Ryan. 1972. *J. Biol. Chem.* **247**:3445.
Mickel, C. E., and J. Standish. 1947. *Minn., Agr. Exp. Sta., Tech. Bull.* **178**.
Mikola, J., and T. M. Enari. 1970. *J. Inst. Brew.* **76**:182.
Mikola, J., and M. Kirsi. 1972. *Acta Chem. Scand.* **26**:787.
Polanowski, A. 1967. *Acta Biochim. Pol.* **14**:389.
Rackis, J. J., and R. L. Anderson. 1964. *Biochem. Biophys. Res. Commun.* **15**:230.
Ryan, C. A. 1966. *Biochemistry* **5**:1592.
Ryan, C. A. 1967. *Anal. Biochem.* **19**:434.
Ryan, C. A. 1973. *Annu. Rev. Plant Physiol.* **24**:173.
Ryan, C. A., and A. K. Balls. 1962. *Proc. Nat. Acad. Sci. U.S.* **48**:1839.
Ryan, C. A., and O. C. Huisman. 1970. *Plant Physiol.* **45**:484.

Ryan, C. A., and L. K. Shumway. 1971. *In* "Proteinase Inhibitors" (H. Fritz and H. Tschesche, eds.), pp. 175–188. de Gruyter, Berlin.

Seidel, D. S., and W. G. Jaffe. 1967. *Enzymologia* **33**:313.

Shain, Y., and A. M. Mayer. 1965. *Physiol. Plant* **18**:853.

Shain, Y., and A. M. Mayer. 1968. *Phytochemistry* **7**:1491.

Vogel, R., I. Trautshold, and E. Werle. 1969. "Natural Proteinase Inhibitors." Academic Press, New York.

Werle, E., and G. Zickgraf-Rudel. 1972. *Z. Klin. Chem. Klin. Biochem.* **10**:140.

REGULATORY CONTROL MECHANISMS
IN ALKALOID BIOSYNTHESIS

HEINZ G. FLOSS, JAMES E. ROBBERS, and PETER F. HEINSTEIN

Department of Medicinal Chemistry and Pharmacognosy,
School of Pharmacy and Pharmacal Sciences,
Purdue University, West Lafayette, Indiana

Introduction

During the early 1950s the field of alkaloid biosynthesis moved from the realms of fruitful and often very farsighted speculation into the phase of experimental investigation. The main driving force for this development, aside from the availability of isotopic tracers, has been the curiosity of organic chemists, who marveled at the ease with which plants seem to carry out complicated synthetic reaction sequences which took the chemists years to work out in the laboratory. Naturally, therefore, the emphasis in these inquiries has been mostly on learning about the chemistry involved in these biosynthetic processes. The questions of why alkaloids are formed and whether or how their formation is controlled have been more the topics of rather opinionated discussions at laboratory coffee breaks or post-seminar get-togethers than of experimental investigations. At the other end of the spectrum, biologically oriented researchers have studied physiological aspects of alkaloid formation in certain plants for very practical reasons, e.g., agronomists in attempts to grow crop plants of low alkaloid content or pharmacognosists in attempts to grow medicinal plants of particularly high alkaloid content. However, although there is a sizeable volume of literature in this area and although these studies have often produced practically useful results, they have, all in all, not contributed very much to our understanding of the regulation of alkaloid biosynthesis. In particular, just like the studies of the organic chemists, these investigations have usually stopped short of isolating the enzymes involved in these processes, an important prerequisite for in-depth regulatory studies. What is conspicuously missing is the involvement of biochemists in this field, only few of whom have apparently considered alkaloid biosynthesis a promising field of study. In all fairness it has to be kept in mind that the isolation of enzymes from plants in general and of enzymes involved in secondary metabolism in particular has been and continues to be a very difficult and frustrating task and only during the last few years has the pace of progress in this area gained momentum. In the alkaloid field this pace is still painfully slow. Five years ago in a review of alkaloid biosynthesis, Leete (1969) wrote: "The enzymes which control the formation of alkaloids are now being investigated by biochemists and plant physiologists. Good methods are available for the isolation of plant enzymes and rapid progress is to be expected in this area." Five years later, this rapid progress still has yet to materialize. Some of the delay is probably due to external factors, and it is to be hoped that Leete's prediction will soon come true.

It seems nevertheless appropriate to discuss at this time some of the work which has been done and which gives some clues as to what types of regulatory mechanisms might be involved in alkaloid synthesis. This could serve as a stimulus for future work and perhaps also as a base line against which to gauge future progress. No attempts have been made to assure complete coverage of the field and the reader will soon discover that much room is given to discussions of the regulation of ergot alkaloid biosynthesis—not only because this happens to be one of the authors' pet topics, but also because there seems to be more basic information available on the regulation of this biosynthesis than on that of most other classes of alkaloids. There is, of course, a good reason for this: ergot alkaloids are formed not only by higher plants but, more prominently, by fungi, which lend themselves much more readily to experimental investigation.

The reader who wishes to familarize himself more thoroughly with the developments in this field is referred to the proceedings of the four symposia on the biochemistry and physiology of alkaloids, which were held in Germany in 1956, 1960, 1965, and 1969 (Mothes and Schütte, 1956; Mothes and Schroeter, 1963; Mothes *et al.*, 1966; 1971). Each of these gives a very good overview of the state of knowledge in this field at the time. For comprehensive reviews of our knowledge of the biosynthetic pathways leading to the various alkaloids the reader may consult reviews by Leete (1967), Spenser (1968), or Mothes and Schütte (1969a).

General Considerations

ALKALOID FORMATION AND GROWTH

It is very difficult to generalize on the relationship between alkaloid synthesis and the overall development of the plant. Usually, alkaloids are synthesized in young tissue, although not necessarily in the actively growing cells. To quote just a few examples, the synthesis of ricinine takes place in germinating seedlings of *Ricinus communis* beginning at day 5. The alkaloid is found mainly in the cotyledons and, in etiolated seedlings, its synthesis terminates as soon as the endosperm is consumed, whereas in illuminated plants it continues (Schiedt *et al.*, 1962). Alkaloids that are synthesized in the roots (nicotine, some *Datura* and *Atropa* alkaloids) are usually formed immediately after germination and their synthesis only continues as long as the growth of the roots continues (Mothes and Schütte, 1969b). On the other hand, there are clear cases in which alkaloid synthesis occurs in older tissues. In lupines, for example, the synthesis of alkaloids does not start until 2 weeks after germination and it continues until the plants start flowering (Wiewiórowski *et al.*, 1966). The onset of flowering seems in many cases to slow down or stop alkaloid synthesis. There are a number of examples of alkaloid synthesis in specialized cells of highly differentiated tissues, e.g., the particulate fraction of the latex of *Papaver somniferum* (Fairbairn *et al.*, 1968). Dedifferentiated tissues usually do not synthesize much alkaloid; this may explain to some extent the poor results obtained so far in attempts to produce alkaloids in plant cell cultures. Seeds are in some cases the site of alkaloid synthesis, as has been shown for the formation of damascenine in *Nigella damascena* (Vislin *et al.*, 1964), but in other cases the alkaloid content of the seeds can also decrease upon ripening (e.g., in *Papaver* or *Nicotiana*).

In the formation of fungal alkaloids, as in other fermentation processes, one can clearly distinguish a distinct growth phase of the organism (termed trophophase), followed by a separate production phase (called idiophase). This has been observed with cultures of various species of *Claviceps* which produce ergot alkaloids (Taber, 1964; Amici *et al.*, 1967a; Kaplan *et al.*, 1969) and with cultures of *Penicillium cyclopium* which produce the benzodiazepines cyclopenin (**1a**) and cyclopenol (**1b**) and, derived from them, the quinoline alkaloids viridicatin (**2a**) and viridicatol (**2b**) (Luckner and Nover, 1971). While in the ergot fungus the formation of conidia usually leads to termination of alkaloid synthesis, the formation of alkaloids in *P. cyclopium* seems to coincide with conidia formation. Luckner and Nover (1971) were able to show that the enzymes responsible for formation of cyclopenin are equally distributed between mycelium and conidia, whereas the enzyme cyclopenase, which catalyzes the conversion $1 \rightarrow 2$ is located exclusively in the spores.

1a R = H   2a R = H
1b R = OH   2b R = OH

TRANSLOCATION AND TRANSPORT

The site of synthesis of an alkaloid is frequently not the site at which it is found in highest concentration. Rather, it seems quite common that alkaloids are removed from their site of synthesis and transported to other sites in neighboring tissues or even other parts of the plant, or that they are excreted into extracellular spaces. The accumulation of alkaloids in the latex in a number of plants (members of Papaveraceae, Lobeliaceae, and Amaryllidaceae) is an example of the latter, although this seems to be less common than previously assumed (Vazujfalvi, 1971). Quite frequently alkaloids accumulate in the vacuoles, presumably by virtue of the fact that these are usually rich in organic acids. This accumulation process in particular is considered to be an important factor in the physiological control of alkaloid formation. There are instances in which alkaloids are accumulated in special cells, e.g., oil cells, as in the case of damascenine, where this has been elegently demonstrated by the use of fluorescence microscopy (Munsche, 1964). Synthesis of alkaloids in the roots and transport to the aerial parts is a well-known process (e.g., in tobacco or in *Datura*), but the reverse process also seems to take place. Deposition of alkaloids synthesized somewhere else than in the seeds is not uncommon. For example, the primary site of synthesis of ergot alkaloids in *Ipomoea* species are the leaves (Mockaitis *et al.*, 1973), but the alkaloids are accumulated in the seeds which themselves cannot synthesize alkaloids (Groeger *et al.*, 1963). In all these cases it is not known whether the transport and the translocation of the alkaloids involves an active transport system in the sense of an energy-dependent, carrier-mediated process, whether it occurs along a concentration gradient, or whether it follows the general nutrient flow.

Some of the fungal alkaloids, e.g., the "water-soluble" ergot alkaloids or cyclopenin and cyclopenol are "excreted" into the culture filtrate. However, this does not necessarily imply an active transport mechanism. The ratio of intracellular to extracellular fluid volumes in such cultures is of the order of 1:25, which means that if the alkaloid equilibrates between intracellular and extracellular fluid, over 95 percent of it will be found in the culture medium. As discussed below, in the case of the clavine alkaloid-producing ergot strain SD 58, our analyses indicate at best a twofold concentration of the alkaloids in the medium over the intracellular fluid (Fig. 16). Some fungal alkaloids, e.g., some of the peptide alkaloids of ergot (Amici *et al.*, 1967a) or the quinoline alkaloids of *P. cyclopium* (Mothes and Schütte, 1969b), are concentrated in the mycelium. It is not clear whether this is due to an active transport process or to other factors such as solubility.

Extensive studies on tryptophan uptake and alkaloid synthesis in *Claviceps* led Teuscher (1965, 1966) to conclude that the ability of strains to produce alkaloids parallels their ability to take up and metabolize tryptophan added to the culture medium. Tryptophan is one of the building blocks of these alkaloids. Comparing several alkaloid-producing and non-producing strains, he found that only the high producers had a pronounced ability to accumulate and/or degrade tryptophan (Teuscher, 1964). This work prompted us to study in some detail the properties of the tryptophan transport system in *Claviceps* strain SD 58, a strain which produces fairly high quantities of clavine alkaloids, mainly elymoclavine. The tryptophan permease system was found (Robertson *et al.*, 1973) to be energy-dependent and shows Michaelis-Menten kinetics (K_m for L-tryptophan 1.6×10^{-4} M); tryptophan transport is linear with time, has a pH optimum of about 5.0 (the initial pH of the culture medium is 5.4), and a temperature optimum of 30°C. Of particular interest was the finding that tryptophan permease activity was very high during the growth phase, with a maximum at around 30–36 hours, but decreased dramatically thereafter (Fig. 1). The uptake of an unrelated amino acid, L-arginine, shows a similar pattern, although the maximum occurs somewhat later (Fig. 1). Since alkaloid synthesis does not start until about day 3–4, there is no positive, but rather at best a negative correlation between tryptophan permease activity and alkaloid synthesis. Comparison of strain SD 58 with the nonproducing strain EK showed that the latter had less tryptophan permease activity (Fig. 2), but the difference was primarily in the growth phase, whereas later in the culture period the permease activity of both strains was quite similar (Robertson, 1971). The tryptophan

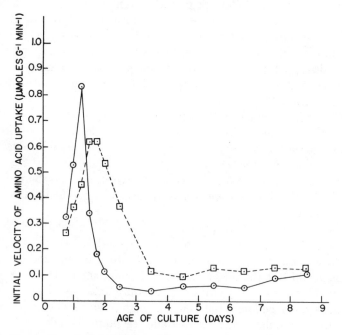

FIG. 1. Comparison of L-tryptophan-β-^{14}C and L-arginine-guanido-^{14}C uptake as function of culture age in *Claviceps* strain SD 58. ⊙, Initial velocity of L-tryptophan uptake; ⊡, initial velocity of L-arginine uptake. (Reproduced from Robertson *et al.*, 1973, with permission of the copyright owner, American Society for Microbiology.)

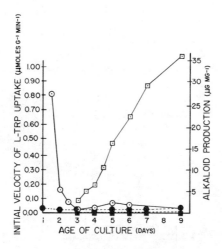

Fig. 2. Alkaloid production and initial velocity of tryptophan transport in *Claviceps* strains SD 58 and EK. ⊙——⊙, Tryptophan uptake in strain SD 58; ▯——▯, alkaloid production in strain SD 58; ●– – –●, tryptophan uptake in strain EK; ■– – –■, alkaloid production in strain EK.

permease activity of strain SD 58 stayed high throughout the culture period if the organism was grown in a medium containing a 10-fold increase in the level of inorganic phosphate, a condition which completely suppresses alkaloid formation (Fig. 3). From these and additional experiments it was concluded that the tryptophan permease activity in this organism is not related in any obvious way to the production of alkaloids.

DEGRADATION

In any closed metabolic system the amount of a given metabolite depends on its rate of formation and its rate of degradation. This is true also for alkaloids. For a long time it was believed that alkaloids were end products of metabolism which were not metabolized any further. Research during the last decade or two has, however, shown that this concept is not generally true, and there are now many well-documented examples of partial or complete metabolic breakdown of alkaloids. The disappearance of alkaloids from ripening seeds in a number of plants has already been mentioned, and this is certainly at least partly due to degradation. The steroid alkaloids present in the fruits of *Solanum dulcamara* are metabolized almost completely as the fruits ripen (Willuhn, 1966). Tracer experiments have clearly demonstrated the metabolism of various alkaloids in their

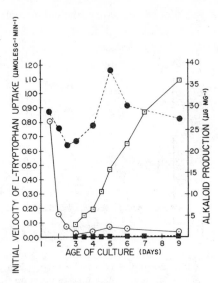

Fig. 3. Alkaloid production and initial velocity of tryptophan transport in *Claviceps* strain SD 58 grown in regular NL 406 and NL 406 high phosphate medium. ⊙——⊙, Tryptophan uptake in NL 406; ⊡——⊡, alkaloid production in NL 406; ●- - -●, tryptophan uptake in NL 406 high phosphate; ■- - -■, alkaloid production NL 406 high phosphate.

producing organisms, e.g., morphine in *Papaver somniferum* (Fairbairn and El-Masry, 1967), caffeine in *Coffea arabica* (Wanner and Kalberer, 1966), or elymoclavine in *Ipomoea rubro-caerulea* (Groeger, 1963). In the latter case, the rapid degradation of the alkaloid in the leaves explains why, despite the fact that they are the primary site of alkaloid synthesis (Mockaitis *et al.*, 1973; Groeger *et al.*, 1963), the leaves have the lowest alkaloid content of all the tissues (Groeger *et al.*, 1963).

The list of examples could be continued to fill many pages; the few given here should suffice to illustrate that degradation of alkaloids is a quite common phenomenon and certainly in many cases a major factor determining the alkaloid content of a plant. If, as the subsequent discussion will illustrate, our knowledge of the regulation of alkaloid synthesis is scanty, then our knowledge of the control of alkaloid breakdown is virtually nonexistent.

GENETIC DETERMINANTS

The fact that alkaloid formation is an enzymatic process leads to the conclusion that alkaloid biosynthesis is under genetic control. While this conclusion may seem obvious, experimental evidence showing the type of genetic control involved is somewhat limited. Breeding studies utilizing alkaloid-bearing plants of the Solanaceae have demonstrated that the F_1 hybrids inherit the alkaloid-producing ability of both parents (Romeike, 1961, 1962, 1966; Hills *et al.*, 1954; Evans *et al.*, 1969). Utilizing biometrical analysis, Solomon and Crane (1970) have recently reevaluated data on *Atropa belladonna* which was reported by Sievers (1915). They came to the conclusion that heredity in alkaloidal phenotypes of *A. belladonna* was under the influence of a polygenic system, a system where the individual effects of genetic factors are equivalent but whose actions intensify each other. Helping to support this concept for alkaloid formation in general are the observations that increased numbers of genes in autotetraploids of species of *Datura* (Rowson, 1945a) and in *A. belladonna* and *Hyoscyamus niger* (Rowson, 1945b) result in increased alkaloid production. Stary (1963) has found using aneuploids of *Datura stramonium* that the 17·18 and 21·22 chromosomes influence the ratio of hyoscine to hyoscyamine as compared to normal diploid plants. A polygenic system may also be important in ergot alkaloid biosynthesis since evidence has been obtained that the heterokaryotic state is associated with alkaloid formation in saprophytic culture (Amici *et al.*, 1967b; Spalla *et al.*, 1969; Spalla, 1972). Using strains of *Claviceps purpurea* which produced peptide alkaloids in submerged culture, homokaryons of each strain which either had lost or had a reduced ability to produce alkaloids were segregated. A recombination of the segregates resulted in regaining a portion of the alkaloid-producing ability of the parent strain from which they had been originally segregated. Apparently as in the case of polyploidy in plants the heterokaryotic state in the fungus ensures an increased diversity of genetic material in the organism allowing for an increased ability for alkaloid formation (see Fig. 4).

One might speculate that if polygenic inheritance is the universal model for control of alkaloid biosynthesis several problems in studying and identifying regulatory mechanisms in alkaloid formation would arise. A polygenic system could involve more than one operon, as evidenced by the work of Stary (1963) that more than one chromosome is involved in the formation of tropane alkaloids in *Datura*. If this were the case, it would be difficult to demonstrate conclusively a repression phenomenon if it were operative. In order to overcome potential problems of this type attempts should be made in regulation studies on alkaloid formation to gather the genetic information for synthesis into haploid forms of the organism. For example, haploid plants have been obtained from pollen grains of various species of *Nicotiana* (Nakata and Tanaka, 1968; Nitsch and

Nitsch, 1969) in our laboratories we have demonstrated that the alkaloid-producing strain of ergot, SD 58, is a monokaryon; hence a haploid strain (Robbers *et al.*, 1972a) (see Fig. 5).

INFLUENCE OF EXTERNAL FACTORS

There is a vast amount of literature describing the influence of various physical and chemical external factors on alkaloid production in plants, a review of which could easily fill a book by itself. These include such variables as climatic conditions, photoperiod, geographic location, fertilization, and the influence of a large array of added chemicals, e.g., plant growth regulators, various inorganic salts, amino acids, carbohydrates, or specific precursors of the particular alkaloids. These factors may have positive or negative effects on alkaloid formation, but except for a few of the more obvious relationships (e.g., compounds affecting the growth of the plant, specific alkaloid precursors) we usually do not have any clue as to how these effects are mediated.

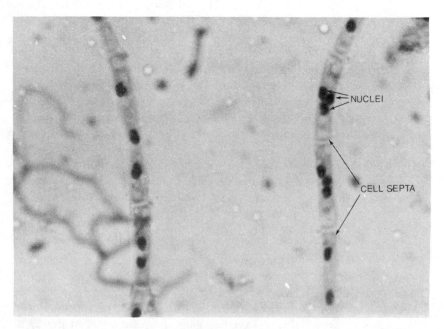

FIG. 4. Hyphal strands showing multinucleated cells of 5-day-old mycelium of an ergotoxine-ergonovine producing, submerged culture of *Claviceps purpurea.* 700×.

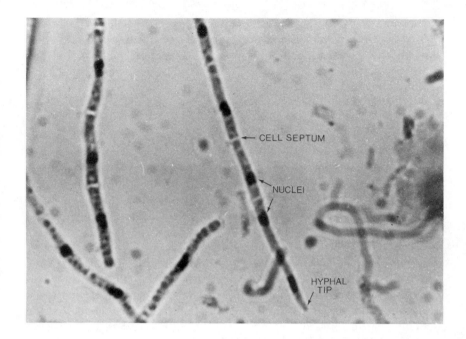

FIG. 5. Hyphal strands showing mononucleated cells of 7-day-old mycelium of *Claviceps* species strain SD 58, which was grown in submerged culture. 700×.

Regulatory Control Mechanisms

REGULATION OF THE SYNTHESIS OF ALKALOID PRECURSORS

One of the ways in which the synthesis of an alkaloid could be regulated is through the synthesis of its primary precursor(s). It is therefore of interest to study the regulation of the synthesis of alkaloid precursors (e.g., the parent amino acids) in the alkaloid-producing organism and to compare it to the same process in other organisms not producing these alkaloids. It is also of interest to determine if the alkaloids have any regulatory effect on the synthesis of their own precursors. Some such studies have been carried out.

The pyridine alkaloids, e.g., anabasine, nicotine, and ricinine, appear to have a special relationship to primary metabolism in that control mechanisms in their formation may involve the regulation of a metabolic pathway leading to a primary metabolite, namely nicotinamide adenine dinucleotide (NAD). In the case of ricinine, Leete and Leitz (1957) demonstrated that the nitrile carbon atom is specifically derived from the carboxyl carbon of nicotinic acid and in this same work it was suggested that before incorporation into ricinine the nicotinic acid is first incorporated into NAD through the pyridine nucleotide cycle. A key intermediate in the biosynthesis of ricinine and also the pyridine nucleotide cycle is quinolinic acid, a compound which can be formed in plants from aspartic acid and glyceraldehyde 3-phosphate or glycerol as postulated from indirect evidence utilizing the incorporation of these compounds into pyridine alkaloids (Yang and Waller, 1965; Fleeker and Byerrum, 1967). Nicotinic acid mononucleotide, an intermediate in the pyridine nucleotide cycle, can be formed from quinolinic acid and phosphoribosyl pyrophosphate without proceeding through nicotinic acid. This precursor role of quinolinic acid in the pyridine nucleotide cycle has been demonstrated in bacteria (Andreoli *et al.*, 1963), in plants (Hadwiger *et al.*, 1963), and in mammalian tissue (Nishizuka and Hayaishi, 1963).

In the biosynthesis of ricinine in young *Ricinus communis* plants, quinolinic acid is a more efficient precursor than nicotinic acid (Yang and Waller, 1965); consequently, Waller *et al.* (1966) explored the possibility that the pyridine nucleotide cycle intermediates might be precursors to ricinine. They found in experiments with a duration of 96 hours that the pyridine ring moieties of quinolinic acid, nicotinic acid mononucleotide, nicotinic acid dinucleotide, and NAD were readily incorporated into ricinine with efficiencies of the same order of magnitude as those of nicotinic acid and nicotinamide; however, the detailed steps in the conversion of any particular pyridine compound to ricinine have not been elucidated.

In addition to the involvement of the pyridine nucleotide cycle in the biosynthesis of ricinine, two other biosynthetic possibilities have been postulated. The first requires nicotinamide as the direct intermediate to ricinine. Waller and Henderson (1961) utilized ^{15}N-labeled nicotinamide and obtained evidence that the cyano nitrogen of ricinine arises from the amide nitrogen of nicotinamide. These results were later reconfirmed with 96-hour experiments (Waller and Yang, 1967). The rationale for this postulation is that nicotinamide cannot be incorporated into ricinine in these experiments via a pyridine nucleotide intermediate because, in order for nicotinamide to be converted to a pyridine nucleotide, it must first be converted to nicotinic acid and the amide nitrogen would be lost in the process.

Another possibility has been suggested by Hiles and Byerrum (1969). They performed competitive feeding experiments using castor bean seedlings and found that exogenous NAD caused an increase in the incorporation of radioactive quinolinic acid into ricinine. If one assumes that NAD is an obligatory intermediate in the biosynthesis of ricinine from quinolinic acid and that exogenous NAD can cross the cell membrane, they reasoned that the exogenously fed NAD should dilute the radioactive quinolinic acid and cause a decrease in the radioactivity incorporated into ricinine. Since they obtained increased incorporation, they postulated that ricinine and NAD are made from quinolinic acid by separate and independent biosynthetic pathways.

What appears to be a complicated story in the biosynthesis of ricinine may be somewhat clarified by considering recent evidence concerning the possible regulation of ricinine biosynthesis. A preliminary report from Waller's laboratory (Johnson and Waller, 1972) revealed a concurrent decrease in the incorporation of radioactivity into ricinine and also into the pyridine nucleotide cycle intermediates in *Ricinus communis* in the presence of a two- to threefold excess of unlabeled ricinine. It was pointed out that this demonstrated the obligatory role of the pyridine nucleotide cycle in the biosynthesis of ricinine from quinolinic acid; however, it also suggests that the biosynthesis of ricinine is regulated at the precursor stage. Supportive evidence to the first conclusion concerning the obligatory role of the pyridine nucleotide cycle comes from Byerrum's laboratory (Mann, 1972). Using etiolated seedlings of *Ricinus communis*, they found a 12-fold increase in ricinine content over a 4-day period and a sixfold increase in quinolinic acid phosphoribosyltransferase activity which preceded the onset of ricinine biosynthesis by one day. This enzyme catalyzes the formation of nicotinic acid mononucleotide from quinolinic acid and phosphoribosyl pyrophosphate and seems to be a point at which regulation may take place in the pyridine nucleotide cycle. In this regard, Gholson *et al.* (1964) showed that a crude beef liver homogenate capable of converting quinolinate to nicotinic acid mononucleotide was inhibited by a 10^{-4} M concentration of NAD and led them to speculate that the level of NAD in liver may serve to regulate its biosynthesis from quinolinate. Furthermore, in a cell-free extract from the alga, *Astasia longa*, it was found that nicotinic acid inhibited quinolinic acid phosphoribosyltransferase activity (Kahn and Blum, 1968). In the case of ricinine biosynthesis, however, apparently the point of regulation is not in the conversion of quinolinic acid to nicotinic acid mononucleotide, since Mann (1972) found no inhibition by ricinine of quinolinic acid phosphoribosyltransferase which had been purified approximately 500-fold from etiolated castor bean endosperm.

Experimental evidence indicates that the formation of the pyridine moiety of the tobacco alkaloids, nicotine and anabasine, closely parallels the biosynthesis of ricinine, namely, an association with the pyridine nucleotide cycle. It has been demonstrated that nicotinic acid can serve as a precursor to nicotine (Dawson *et al.*, 1960) and to anabasine (Solt *et al.*, 1960). In the case of nicotine, quinolinic acid is an efficient precursor of nicotine in whole tobacco plants (Yang *et al.*, 1965) and NAD is incorporated into nicotine with an efficiency similar to quinolinic acid and nicotinic acid (Frost *et al.*, 1967). Finally, similar to the results for ricinine, Mann (1972) discovered elevated levels of quinolinic acid phosphoribosyltransferase in *Nicotiana rustica* roots which synthesized nicotine. If indeed, as it now appears, the biosynthesis of ricinine is regulated at the level of precursor formation, one might speculate that the formation of nicotine and anabasine may also be regulated in a similar way because of the close similarity in the biosynthesis of the pyridine moiety of these compounds. The prior discussion is summarized in Fig. 6.

Another system in which the regulation of alkaloid precursor synthesis has been studied to some extent is the ergot fungus. The alkaloids produced by this fungus are derived from tryptophan, mevalonic acid, and the methyl group of methionine (Weygand and Floss, 1963) (Fig. 7). Experiments in our laboratory (see below) and by others (Arcamone *et al.*, 1962) have shown that various tryptophan analogs, some of which in other

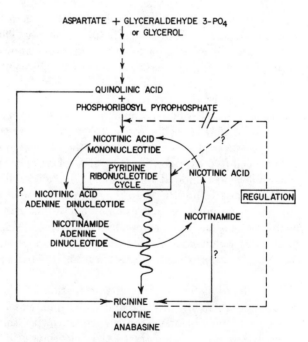

Fig. 6. Scheme for the biosynthesis of the pyridine moiety of the pyridine alkaloids.

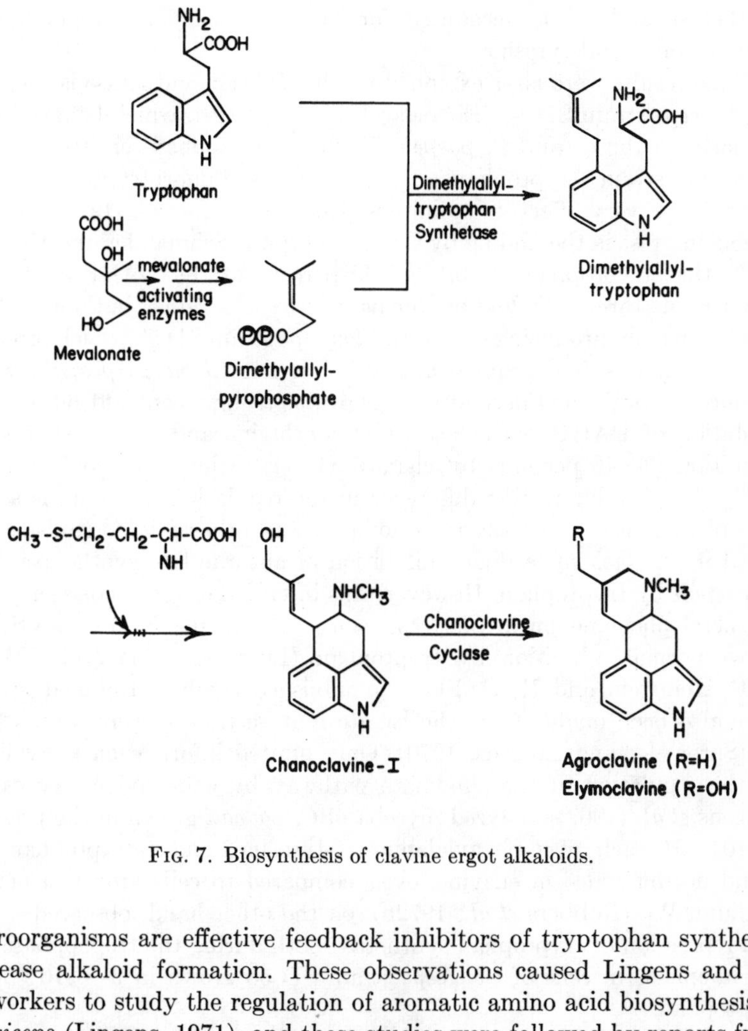

FIG. 7. Biosynthesis of clavine ergot alkaloids.

microorganisms are effective feedback inhibitors of tryptophan synthesis, increase alkaloid formation. These observations caused Lingens and his co-workers to study the regulation of aromatic amino acid biosynthesis in *Claviceps* (Lingens, 1971), and these studies were followed by reports from other laboratories. Lingens *et al.* (1967) first studied a number of the shikimate pathway enzymes from alkaloid-producing mycelia of *Claviceps paspali* and compared their allosteric regulation to that of the same enzymes in other microorganisms. Chorismate mutase, prephenate de-hydrogenase, and prephenate dehydratase showed no unusual behavior, but in the case of 3-deoxy-D-arabinoheptulosonic acid-7-phosphate (DAHP) synthetase, the first enzyme of the pathway, the isoenzyme which was inhibited by L-tryptophan accounted for 60 percent of the total activity. In contrast, in most other microorganisms 85 percent of the DAHP synthetase activity is accounted for by the isoenzymes responding to phenylalanine and tyrosine.

These results were later extended to the DAHP synthetase isoenzymes of *Claviceps* strain SD 58 (Eberspaecher *et al.*, 1970), which behaved very similarly to those from *C. paspali*. With the trytophan-sensitive enzyme from this strain, tryptophan analogs, such as 5-fluorotryptophan, were found to be very effective inhibitors (Eberspaecher *et al.*, 1970). Of particular interest is the finding by Groeger's group (Schmauder and Groeger, 1973) that tryptophan inhibits 45–60 percent of the DAHP synthetase activity in three alkaloid-producing *Claviceps* strains, but only 20–25 percent in a nonproducing strain. In *Claviceps* strain SD 58, which produces elymoclavine as the terminal alkaloid, but not *Claviceps paspali*, which produces lysergic acid derivatives, there was a significant (40–60 percent) inhibition of DAHP synthetase by elymoclavine and a somewhat lesser inhibition (20–40 percent) by chanoclavine-I (Schmauder and Groeger, 1973). Another interesting difference in the regulation of aromatic amino acid biosynthesis in *Claviceps* as compared to other microorganisms was noted in the lack of feedback inhibition of anthranilate synthetase from *C. paspali* by tryptophan. However, this insensitivity does not seem to be a general phenomenon in *Claviceps*, since the enzyme from strain SD 58 shows normal inhibition by tryptophan (Lingens, 1971; J. D. Mann, P. F. Heinstein and H. G. Floss, unpublished results). Detailed studies have also been made of the chorismate mutase from *Claviceps* strain SD 58 (Sproessler and Lingens, 1970). Only limited information is available on the regulation of the shikimate pathway by end-product repression. Lingens *et al.* (1967) analyzed mycelia of *C. paspali* grown in the presence of 10^{-3} *M* each of L-phenylalanine, L-tyrosine, and L-tryptophan and found no difference in enzyme levels compared to cells grown in normal medium. We (Robbers *et al.*, 1972b), on the other hand, observed a five-fold repression of tryptophan synthetase both during the trophophase and the idiophase in cells of *Claviceps* strain SD 58 grown in 2×10^{-3} *M* L-tryptophan.

Some efforts have been made to determine the time course of the appearance and disappearance of the tryptophan biosynthetic enzymes in *Claviceps* during the culture period. Several groups have reported an increase in tryptophan synthetase activity at the beginning of the idiophase in alkaloid-producing strains (Schmauder and Groeger, 1973; Robbers *et al.*, 1972b; Řeháček *et al.*, 1971). In two strains the activity of the enzyme reaches a maximum at the time of most rapid alkaloid formation and then decreases very dramatically (Schmauder and Groeger, 1973). The same time course is seen for anthranilate synthetase (Schmauder and Groeger,

1973; J. D. Mann, P. F. Heinstein, and H. G. Floss, unpublished results). DAHP synthetase activity shows an earlier increase and a much slower decline (Schmauder and Groeger, 1973).

INDUCTION OF ALKALOID SYNTHESIS

As mentioned earlier, in many systems the synthesis of alkaloids does not take place in actively growing cells, but rather only after the growth phase is over. This is particularly true for alkaloid synthesis in microorganisms. This could either be due to lack of precursors during the growth phase or, much more likely, because the enzymes involved in alkaloid formation are repressed in growing cells. Their synthesis and with that the beginning of alkaloid formation would then involve a derepression or induction process. To our knowledge, no unequivocal proof for such a process is presently available for any alkaloid-synthesizing system, but there is a certain amount of circumstantial evidence.

It has been known for quite some time (cf. Tyler, 1961) that the addition of tryptophan to cultures of the ergot fungus in many cases increased the alkaloid yield, and this had been attributed to its role as an alkaloid precursor. In 1964, however, we (Floss and Mothes, 1964) made some observations which pointed to an additional role of tryptophan in the synthesis of ergot alkaloids. It was noted (Table 1) that the addition of L-tryptophan to cultures of ergot strain SD 58, after the end of the growth phase and during the alkaloid production phase, did not increase alkaloid concentration significantly, unless a very large amount of tryptophan was added, but that a consistent increase in alkaloid yeild was obtained when tryptophan was added at the beginning of the growth phase. Yet this was not due to the formation of more mycelium, but rather, the same amount of mycelium synthesized more alkaloid. It was then found that a similar stimulation of alkaloid synthesis could be obtained by adding various tryptophan analogs to the cultures (Table 2) and, furthermore, that mycelium grown in the presence of tryptophan or tryptophan analogs retained an ability to produce more alkaloid than the controls even after it had been transferred into new culture medium not containing the effectors (Table 3). Since these analogs did not give rise to ergot alkaloid analogs and since, at least in the cases checked by radioactive labeling, they were not incorporated into the ergoline ring system, their stimulation of alkaloid synthesis cannot be explained by a precursor effect. On the basis of these results we (Floss and Mothes, 1964) proposed that tryptophan has a dual role in ergoline biosynthesis: it is a precursor of the alkaloids and it is also in some way involved in the induction of the alkaloid-synthesizing enzymes.

TABLE 1

<small>THE EFFECTS OF EARLY AND LATE ADDITION OF TRYPTOPHAN ON THE
FORMATION OF ALKALOIDS</small>

			Alkaloid production		
Culture[a]	Trypto-phan (mM)	Control (mg/liter)	With trypto-phan (mg/liter)	Percent of control
1. Addition after growth phase	M 24	1	880	783	89
			880	865	98
			847	830	98
			710	653	92
			870	925	106
	M 33	2.5	645	641	100
	M 52	1	973	990	102
		10	973	1432	147
2. Addition before growth phase	M 34	2	604	876	145
				1048	174
	M 49	2	500	805	162

[a] M 24 = surface cultures, 100 ml, culture period: 42 days, addition of tryptophan on the 16th day. M 33 = shake culture, 100 ml, culture period: 20 days, addition of tryptophan on the 10th day. M 34 and M 49 = shake cultures, 25 ml, culture period: 27 and 22 days, resp., addition of tryptophan before inoculation. M 52 = surface culture, 100 ml, culture period: 58 days, addition of tryptophan on the 26th day. Reproduced from Floss and Mothes (1964) with permission of the copyright owner, Springer-Verlag.

This hypothesis has since received support by work from several laboratories including our own.

Bu'Lock and Barr (1968),working with *Claviceps purpurea* strain PRL-1980 in a fully synthetic medium, observed a particularly clear-cut dependence of alkaloid production on early addition of tryptophan to the cultures. They also found that, in this strain, the addition of L-tryptophan induced the formation of up to twice as much alkaloid as corresponded to the amount of tryptophan added, whereas in the absence of tryptophan virtually no alkaloid was formed. In such tryptophan-supplemented cultures the alkaloid production curve typically showed two inflection points at days 6 and 10, corresponding to minima in the curves for the rate of alkaloid synthesis. The second differential of the alkaloid production curves (d^2Alk/dt^2), which would indicate the rate of the appearance and disappearance of an enzyme(s) limiting the rate of alkaloid synthesis, closely paralleled the experimental curve for internal tryptophan concentration, suggesting that "the rate of synthesis of this enzyme is in direct response to the tryptophan within the mycelium." The fluctuations in the rate of alkaloid synthesis were explained by the operation of a feedback control system involving "hunting" (Bu'Lock and Barr, 1968). The stimulation of ergot alkaloid synthesis by addition of tryptophan at the beginning of the growth phase (within one day after inoculation), but not at a later time, was also observed by Vining (1970) using *Claviceps* strain HLX 123, which, like strain SD 58, had been isolated from *Pennisetum* grass. Vining's experiments also support the idea of a regulatory function of tryptophan.

Our own further studies first of all led to a system in which the induction of alkaloid synthesis could be more clearly demonstrated. The previous

TABLE 2

THE EFFECT OF TRYPTOPHAN ANALOGS ON THE PRODUCTION
OF ALKALOIDS

Additions	Alkaloids (%)		Mycelium dry weight (mg/culture)	
	M 34[a]	M 49[a]	M 34[a]	M 49[a]
Control[b]	100	100	100	100
L-Tryptophan	145	161	97	95
	174	—	—	—
4-Methyltryptophan	—	126	—	98
5-Methyltryptophan	82	161	95	92
6-Methyltryptophan	—	158	—	86
	—	176	—	—
7-Methyltryptophan	129	110	85	83
	—	—	79	—
1-*N*-Methyltryptophan	134	106	84	85
Homotryptophan	—	110	—	49
Bishomotryptophan	—	125	—	94

[a] M 34 and M 49 were shake cultures, 25 ml each; the tryptophan analogs were added before inoculation at a concentration of 2 m*M*. Culture period was 27 days for M 34 and 22 days for M 49.

[b] The control values were 15.1 mg of alkaloids per culture for M 34 and 12.5 mg per culture for M 49. Reproduced from Floss and Mothes (1964) with permission of the copyright owner, Springer-Verlag.

TABLE 3

The Effect of Preculture of the Mycelium with Tryptophan,
7-Methyltryptophan, and 1-N-Methyltryptophan in the
Medium on the Production of Alkaloids[a]

	Control		Tryptophan		7-Methyl-tryptophan		1-N-Methyl-tryptophan	
	mg	%	mg	%	mg	%	mg	%
Preculture: Alkaloids/1	264	100	409	155	291	110	295	112
Mycelium	1322	—	1388	—	1148	—	1241	—
Main culture: Alkaloids/1	351	100	474	135	422	120	415	118
Cultured mycelium	1065	—	1079	—	856	—	898	—
Harvested mycelium	2284	—	2248	—	2117	—	2119	—
Alkaloids/gm cultured mycelium	33	100	44	134	49	149	46	139
Alkaloids/gm harvested mycelium	15	100	21	136	20	130	19	124

[a] Shake cultures were used (100 ml); compounds were added before inoculation at a concentration of 2 mM. The preculture period lasted for 5 days; thereafter the mycelium was collected, recultured, without additions, and harvested after 12 days. Reproduced from Floss and Mothes (1964) with permission of the copyright owner, Springer-Verlag.

induction experiments with *Claviceps* strain SD 58 were carried out using a medium (NL 406) which in itself allows fairly high alkaloid production, and the relative increases obtained by adding tryptophan or tryptophan analogs were therefore only small. Omitting the yeast extract, the only nondefined ingredient, from NL 406 reduces alkaloid production per flask by a factor of 3 (Fig. 8a). This is mainly due to a decrease in the amount of mycelium formed (Fig. 8c); the ability of the mycelium to synthesize alkaloids is only slightly impaired (Fig. 8b). Addition of L-tryptophan to such cultures, while having little effect on the amount of mycelium formed, restores alkaloid synthesis to its original level, because the same amount of mycelium now synthesizes considerably more alkaloid (Fig. 8a–c) (D. K. Onderka and H. G. Floss, unpublished results). Incidentally, the two inflection points in the alkaloid production curve of a tryptophan-supplemented culture, which Bu'Lock and Barr (1968) noted with strain PRL-1980, can also be seen in Fig. 8a. In such a system the synthesis of alkaloid can be stimulated much more significantly not only with tryptophan, but also with analogs like 5-methyltryptophan (Fig. 9) (Robbers and Floss, 1970). 5-Methyltryptophan was not as good an

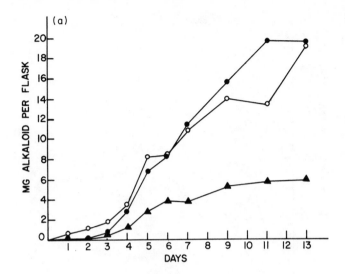

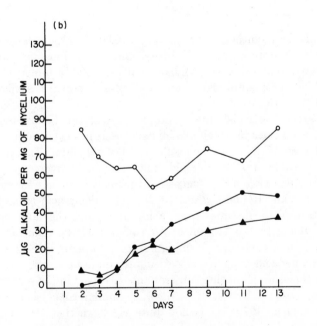

FIG. 8. See pp. 161 and 162. (a) Total, and (b) relative alkaloid production, mycelial growth (c) of *Claviceps* strain SD 58 in synthetic and semisynthetic media. ●, NL 406; ▲, NL 406 minus yeast extract; ○, NL 406 minus yeast extract plus L-tryptophan.

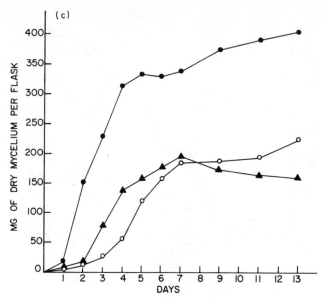

FIG 8c . See p. 161.

inducer of alkaloid synthesis as tryptophan itself. However, later experiments (V. M. Krupinski, J. E. Robbers, and H. G. Floss, unpublished results) showed that other tryptophan analogs, notably 6-methyltryptophan and thiotryptophan(β-(1-benzothien-3-yl)-alanine), are as effective as tryptophan (Fig. 10).

Another system in which the tryptophan effect can be demonstrated very clearly is cultures in which alkaloid synthesis has been suppressed by increased levels of inorganic phosphate (deWaart and Taber, 1960). The phosphate inhibition of alkaloid synthesis can be overcome by exogenous tryptophan (Fig. 11b), suggesting that it is in some way mediated through tryotphan (Robbers *et al.*, 1972b). Mevalonic acid alone has no effect and tryotphan plus mevalonic acid have about the same effect as tryptophan alone. Figure 11a shows for comparison the influence of the same additives in a medium containing normal phosphate levels (NL 406). Again, mevalonic acid is ineffective or slightly inhibitory, suggesting that mevalonate is neither a limiting substrate nor involved in the regulation of this biosynthesis (Robbers *et al.*, 1972b). A similar conclusion has been reached by Vining (1970). It is not quite clear whether the phosphate inhibition of alkaloid synthesis can be overcome by tryptophan analogs. We (Robbers *et al.*, 1972b) obtained very little restoration of alkaloid synthesis with 5-methyltryptophan, but Groeger and co-workers (Erge *et al.*, 1973) were able to induce alkaloid synthesis in phosphate-inhibited cultures with tryptophan or 5-methyltryptophan.

If, as we postulate, the formation of the alkaloid-synthesizing enzymes is induced by tryptophan and if this process is a normal step in the initiation of alkaloid synthesis in *Claviceps*, a number of predictions can be made which can be examined experimentally. First of all, this would require protein synthesis to take place during the transition from growth phase to alkaloid production phase and most likely also further on in the alkaloid production phase. Three groups, including our own (Kaplan *et al.*, 1969; Bu'Lock and Barr, 1968; Rothe and Fritsche, 1967), have shown that this is indeed the case. Moreover, Bu'Lock and Barr (1968) have demonstrated that alkaloid synthesis ceases within 24 hours after inhibition of protein synthesis by cycloheximide. Secondly, if tryptophan induces alkaloid

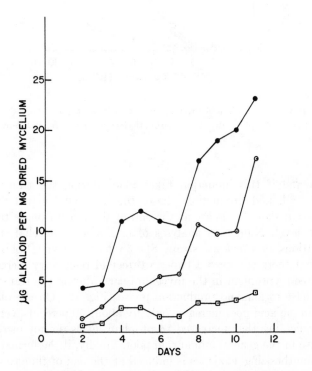

Fig. 9. Time-course study comparing the influence of tryptophan and 5-methyltryptophan on alkaloid production in *Claviceps* strain SD 58. ●, NL 406 minus yeast extract plus D,L-tryptophan; ⊙, NL 406 minus yeast extract plus D,L-5-CH₃-tryptophan; ⊡, NL 406 minus yeast extract. [Reproduced from Robbers and Floss (1970), with permission of the copyright owner, the American Pharmaceutical Association.]

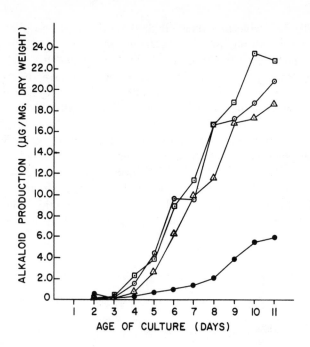

F<small>IG</small>. 10. Induction of alkaloid production in *Claviceps* strain SD 58. ⊙, D,L-trypto-phan; ▢, D,L-thiotryptophan; △, D,L-6-methyltryptophan; ●, D,L-5-methyltrypto-phan.

synthesis and if this induction takes place during the transition from growth to alkaloid production phase, the intracellular free tryptophan concentration should increase at the end of the growth phase to exceed a triggering level. Numerous analyses of endogenous mycelial tryptophan concentrations in *Claviceps* strain SD 58 have shown (Robbers *et al.*, 1972b) that there is indeed a two- to threefold temporary increase of the level of free tryptophan in the mycelium early during the transition from growth phase to alkaloid production phase (Fig. 12). On the other hand, the free amino acid pool increases only by about 50 percent over the same period, indicating that increased tryptophan levels do not merely reflect an increase in the general amino acid pool. Finally, if the formation of the alkaloid-synthesizing enzymes is induced at the end of the growth phase, these enzymes should be absent in growth phase mycelia. Unfortunately, only a few of the enzymes involved in this biosynthesis have so far been isolated. However, the two for which the time course has been studied show the expected behavior. Figure 13 shows the time course of the appearance and disappearance of dimethylallyl pyrophosphate:tryptophan dimethyl-allyltransferase (DMAT synthetase), the first enzyme in the pathway (Heinstein *et al.*, 1971). The enzyme level is very low during the growth phase (the upper dotted line shows the measured activities, the lower one represents the true values after correction for enzyme activity introduced with the inoculum) and increases very markedly during the transition period, preceding the appearance of alkaloids by about $\frac{1}{2}$–1 day. Interestingly, there is a very pronounced decrease of enzyme activity beginning on day 10 when the alkaloid concentration has reached its maximum level. This increase and decrease is reminiscent of the patterns observed for tryptophan synthetase and anthranilate synthetase (see the preceding section). An almost identical time course was reported by Groeger's group for chanoclavine cyclase, an enzyme or enzyme complex functioning somewhat later in the ergoline biosynthetic pathway (Fig. 14) (Erge *et al.*, 1973). These authors also found that cell-free extracts of mycelia of strain SD 58 grown in high-phosphate medium, which did not produce alkaloid, had no chanoclavine cyclase activity, but when alkaloid synthesis was induced with tryptophan or 5-methyltryptophan, cyclase activity was found.

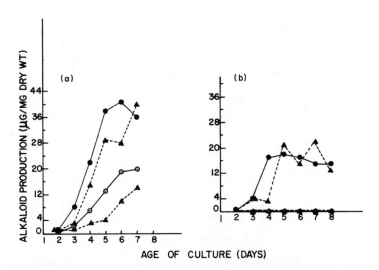

FIG. 11. Comparison of the influence of L-tryptophan and DL-mevalonic acid on alkaloid production in *Claviceps* strain SD 58; (a) in medium containing "normal" (0.1 gm/liter) levels of KH_2PO_4; (b) in medium containing high (1.1 gm/liter) levels of KH_2PO_4. ⊙, No addition; ●, L-tryptophan; △, D,L-mevalonic acid; ▲, L-tryptophan plus D,L-mevalonic acid. (Reproduced from Robbers *et al.*, 1972b, with permission of the copyright owner, American Society of Microbiology.)

All these experiments are in agreement with and support the idea that the formation of the alkaloid-synthezing enzymes is subject to an induction or derepression process and that tryptophan is in some way involved in this process. However, one has to keep in mind that all this does not constitute definite proof for such a process and that most certainly we do not yet fully understand exactly what is going on.

Much less is known about the dynamics of the enzymes involved in other alkaloid biosyntheses and whether induction or depression mechanisms play a role in other systems as well. For one thing, the number of enzymes (involved specifically in other alkaloid biosyntheses) which have been isolated to date is still very small. Few time-course studies have been performed. However, it has been found, for example that the enzyme cyclopenase from *Penicillium cyclopium*, (see Alkaloid Formation and Growth) appears only

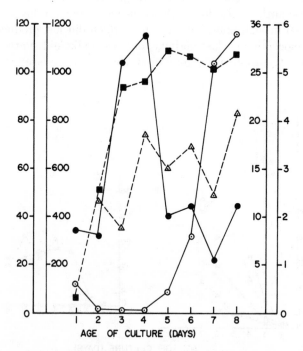

FIG. 12. Comparison of mycelial growth, free endogenous tryptophan, free endogenous amino acids, and alkaloid production in *Claviceps* strain SD 58. △, Free amino acids (μmole/gm dry mycelium); ■, mycelial dry weight (mg); ⊙, alkaloid production (μg/mg dry mycelium); ●, free tryptophan (μmole/gm dry mycelium. (Reproduced from Robbers *et al.*, 1972b, with permission of the copyright owner, American Society of Microbiology.)

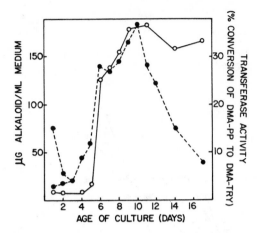

Fig. 13. Time course of the appearance of dimethylallylpyrophosphate: tryptophan dimethylallyltransferase activity and the accumulation of alkaloids in shake cultures of *Claviceps* strain SD 58. ○, Alkaloids; ●, enzyme activity. (Reproduced from Heinstein *et al.*, 1971.)

after the end of the growth phase, simultaneously with the beginning of alkaloid and conidiospore formation (Luckner and Nover, 1971). A rather extensive study has been made of the enzyme tyramine methylpherase (Mann and Mudd, 1963), which is involved in the conversion of tyramine to hordenine in roots of germinating barley seedlings. Mudd and co-workers (Mann *et al.*, 1963) found that the enzyme level rapidly increased during the first 3 days of germination to reach a maximum at about day 5–6, followed by a gradual decline over the next 20-day period. Appearance of the mono- and dimethylated tyramine followed the same time course except for a slight lag. Embryos grown in culture showed a very similar pattern. The enzyme was almost evenly distributed along the roots, with possibly a slightly higher activity in the older segments. Appearance of the enzyme during germination seems to involve *de novo* synthesis, since it was selectively inhibited by puromycin and by analogs of valine, lysine, or phenylalanine. The latter inhibition was reversed by the corresponding amino acids. In cultured embryos, formation of the enzyme was stimulated by casein hydrolysate as well as by several of the amino acids present in casein hydrolysate, lysine being the most effective. It was assumed that this was due to correction of slight dietary deficiencies during the early growth of embryos (Mann *et al.*, 1963).

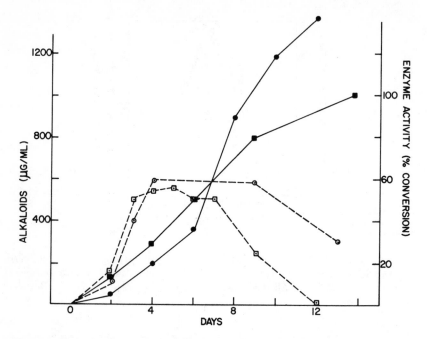

Fig. 14. Chanoclavine-I cyclase activity and alkaloid formation during the culture period in *Claviceps* strains SD 58 and PEPTY 695. ●, Alkaloids PEPTY 695; ⊙, enzyme PEPTY 695; ■, alkaloids SD 58; ☐, enzyme SD 58. (Reproduced from Erge *et al.*, 1973, with permission of the author and the copyright owner, VEB Gustav Fischer Verlag.)

FEEDBACK REGULATION OF ALKALOID SYNTHESIS

An important regulatory mechanism in primary metabolism is the control which the end product of a biosynthetic pathway exerts over its own rate of formation. This can involve allosteric inhibition of a key enzyme, usually the first specific enzyme in the pathway (feedback inhibition), and/or repression of the synthesis of the biosynthetic enzymes involved (end-product repression). Although this has not been studied very extensively, there are some indications that feedback regulation mechanisms may also be operative in alkaloid biosynthesis. Gross *et al.* (1970) have shown that the formation of gramine from β-^{14}C-tryptophan in *Hordeum vulgare* seedlings is decreased by the simultaneous feeding of gramine over and above the normal dilution effect. Furthermore, feeding of increasing amounts of gramine-(methylene-^{14}C) to seedlings did not increase gramine concentration in the young plant above normal levels. However, the specific radioactivity of the gramine in the same plants steadily increased. Both experiments imply either that gramine feedback regulates its own formation from tryptophan or that gramine activates or induces its degradation whenever a sufficient concentration is reached in the plants. Proof of a feedback regulation mechanism, however, awaits isolation of the enzymes catalyzing the reactions from tryptophan to gramine and determination of their susceptibility to increasing concentrations of gramine.

The first enzyme in the pathway leading from tryptophan to N,N-dimethyltryptamine in *Phalaris tuberosa* is tryptophan decarboxylase (Baxter and Slaytor, 1972a). The activity of this enzyme rose sharply during the 4th day after germination and decreased after day 5 to 1 percent at day 19. A number of alkaloids were found to inhibit the enzyme, including N,N-dimethyltryptamine itself at a concentration of 1 mM, which appears to be comparable to the physiological concentration of 1 μmole of N,N-dimethyltryptamine per gram fresh weight (Baxter and Slaytor, 1972b). Therefore, it was reasonable to postulate that N,N-dimethyltryptamine regulates its own formation through a feedback mechanism. *In vivo* experiments showed (Baxter and Slaytor, 1972b) that ^{14}C-tryptophan incorporation into N-methyl- and N,N-dimethyltryptamine reached a peak at day 1 after germination and the amount of ^{14}C in these tryptamines then rapidly decreased to about 15 percent at day 2, suggesting rapid turnover of the alkaloids as another factor controlling the stationary alkaloid concentration. It is difficult to evaluate the relative importance of these two control mechanisms *in vivo*.

The three enzymes which catalyze the first three reactions in the formation of nicotine from ornithine (Fig. 15) have been isolated from *Nicotiana tabacum* (Mizusaki *et al.*, 1973). Upon decapitation of 4-week-old plants, the three enzyme activities increased 2- to 10-fold, reaching maximum activity 24 hours after decapitation, and then decreased again during the following 1–2 days. The accumulation of nicotine in the roots of the de-

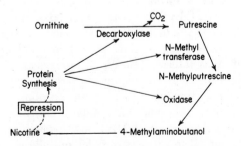

Fig. 15. Nicotine biosynthesis in *Nicotiana tabacum*.

capitated plants followed a similar pattern as the three enzyme activities. Feeding of nicotine to decapitated plants prevented the observed increase in enzyme activities. However, in *in vitro* experiments neither of the three enzymes was inhibited by nicotine (Mizusaki *et al.*, 1972). It therefore appears that nicotine acts as a repressor of the synthesis of the three enzymes. Indoleacetic acid was found (Mizusaki *et al.*, 1973) to stimulate the synthesis of the three enzymes at low concentrations (about 2.5 μM). From these experiments the regulation of the synthesis of nicotine from ornithine was proposed to occur in the following manner: decapitation raises the auxin concentration in the roots sufficiently to stimulate the synthesis of the three enzymes, ornithine decarboxylase, putrescine *N*-methyltransferase, and *N*-methylputrescine oxidase. This increase in enzyme concentrations causes a rise in nicotine concentration in the roots from 0.2 to 0.5 mM, which is sufficiently high to repress the formation of the three enzymes at least 50 percent.

The isolation of the first enzyme, DMAT synthetase, in the pathway leading to the ergot alkaloids (Heinstein *et al.*, 1971) permitted the study of a classical feedback inhibition mechanism. It was shown (S. L. Lee, H. G. Floss, and P. F. Heinstein, unpublished results) that agroclavine and elymoclavine, the end- products of the pathway in this organism, inhibited DMAT synthetase from *Claviceps* strain SD 58 90 percent and 70 percent, respectively, at a concentration of 3 mM. A slight inhibition (27 percent) by 1 mM elymoclavine was also observed for a later enzyme in the pathway, the chanoclavine-I cyclase (Erge *et al.*, 1973). These concentrations are well within the range of the normal alkaloid concentrations in the culture filtrate (250–750 μg/ml = 1–3 mM). The intracellular alkaloid concentrations in mycelia from shake cultures of strain SD 58 were found to be slightly lower (Fig. 16) than those in the culture medium (V. M. Krupinski, J. E. Robbers, and H. G. Floss, unpublished results). (In evaluating the data in Fig. 16, one has to keep in mind that the internal concentrations measured represent minimum values, because some alkaloid may have been lost from the cells during the washing process.) Thus it appears that the inhibition of DMAT synthetase and possibly other enzymes in the pathway by the end product is the physiological mechanism by which alkaloid synthesis is terminated once a certain alkaloid concentration has been reached. This interpretation is in agreement with early experimental observations that a given amount of mycelium can synthesize large amounts of alkaloid continuously if the culture medium is periodically replaced by new medium (Abe and Yamatodani, 1964; Groeger and Erge, 1961). Whether the decrease in DMAT synthetase and chanoclavine-I cyclase activity observed in extracts of older mycelia is due to inhibition

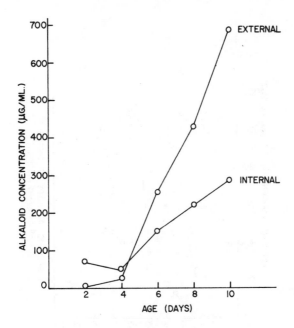

Fig. 16. Intracellular and extracellular alkaloid concentration in shake cultures of *Claviceps* strain SD 58.

of these enzymes by the accumulated alkaloid or to actual decrease in enzyme level remains to be determined. An interesting phenomenon was observed when the intracellular alkaloid concentrations were measured in mycelia obtained from stationary cultures of strain SD 58. These show pronounced fluctuations (Fig. 17) which are reflected in inflection points in the curve representing the alkaloid concentration in the mycelium. As one possibility to rationalize this observation we venture the following explanation: In the stationary cultures, diffusion of the alkaloid out of the mycelium into the medium is a slow process (as is evidenced, for example, by the much higher intracellular concentration). Synthesis of alkaloid and slow diffusion leads to accumulation of alkaloid in the cells. When it reaches concentrations which inhibit the biosynthetic enzymes, *de novo* synthesis slows down or stops until enough alkaloid has diffused out of the cells to lower the intracellular concentration sufficiently to reinitiate synthesis. We suggest that this process repeats itself to give an oscillating system which eventually comes to a stop when the extracellular alkaloid concentration becomes too high to allow further synthesis. Whether this explanation is correct remains, of course, to be determined. As an alterna-

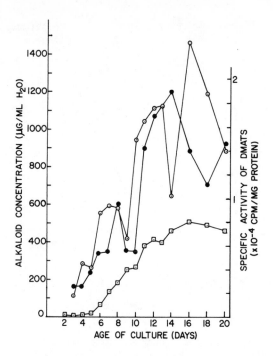

FIG. 17. Relationship of dimethylallyltryptophan synthetase, intracellular, and extracellular alkaloid concentration in stationary cultures of *Claviceps* strain SD 58. ⊙, Internal; □, external; ●, DMAT synthetase.

tive, this observation may simply reflect the possibility that the growth of the cells in a stationary culture is less synchronized than in a shake culture.

INDUCTION OF ALKALOID DEGRADATION

As pointed out above, degradation of alkaloids in the producing organism is a well-documented phenomenon. In other catabolic pathways, synthesis of the enzymes involved is often induced by the substrate and one might expect a similar situation to exist in alkaloid metabolism. Although one can cite a variety of circumstantial evidence in support of it, this idea has so far not been given a thorough test. It is seen in a number of alkaloid-synthesizing plants, particularly in young seedlings, that the alkaloid content increases rapidly initially to reach a plateau and then decreases again, which would be expected for a situation in which a degradative system is induced at a certain alkaloid level. Schütte and his group (Gross *et al.*, 1970) found that feeding increasing amounts of gramine-(methylene-^{14}C) produced increasing specific radioactivities of the cellular gramine, but did not increase the total amount of gramine present in the plant. This suggests that at a certain concentration gramine induces its own degradation, although the situation is not clear-cut, because it is not known what the degradation products of gramine are. If the degradation process is a simple reversal of the biosynthetic pathway, as some work suggests (Digenis, 1969), then the balance of the cellular concentrations of gramine and tryptophan could force the equilibrium either to the left (tryptophan formation from gramine) or to the right (gramine formation from tryptophan).

Other Control Mechanisms Affecting Alkaloid Synthesis

It seems likely that the regulatory mechanisms discussed above are not the only ones controlling alkaloid synthesis, and that as research in this area gains momentum we will undoubtedly uncover additional ways in which the plant controls the flow of metabolites in secondary biosynthetic pathways and the concentration of alkaloids in the tissues. For example, another mechanism for exerting selective control over a secondary biosynthetic pathway, which is just coming to light, results from the evolution of multienzyme complexes, which carry all the enzymes of a particular biosynthetic pathway in one aggregate. This provides a means of channeling a substrate specifically into a particular reaction sequence and prevents intermediates from being drained off into other directions. The recent isolation from sorghum seedlings of a particulate fraction containing all the enzymes necessary to convert tyrosine into a hydroxynitrile corresponding to the cyanogenic glycoside dhurrin (McFarlane *et al.*, 1973) provides an excellent example for the occurrence of this type of control in plants. It would certainly be worthwhile to look for such multienzyme systems in alkaloid biosynthesis. Compartmentalization of substrates and intermediates by confining the enzymes of the reaction sequence to a particular organelle, e.g., the chloroplast, is another means of control which most likely will turn out to be of importance in alkaloid biosynthesis. There are already a number of indications for the occurrence of secondary biosynthetic pathways in chloroplasts and, at least in the case of sparteine, a conspicuously high concentration of alkaloid has been found in the neighborhood of the chloroplasts (White and Spencer, 1964). Another possible regulatory mechanism in alkaloid biosynthesis, about which so far we know virtually nothing, is the removal of a catabolite repression, which may lead to the synthesis of secondary biosynthetic enzymes as the cell comes out of the phase of most active growth.

Physiological and Evolutionary Significance of Alkaloids

Before closing let us consider the physiological and evolutionary significance of alkaloids in the light of the previous discussions on the regulation of alkaloid synthesis. Many hypotheses have been presented to rationalize the formation of alkaloids in plants. Mothes in his Paul-Karrer lecture (Mothes, 1969) pictured alkaloids as metabolic excretion products without physiological function in the plant, the multitude of chemical structures formed being a manifestation of the many chemical capabilities of plants. In earlier discussions, he used the term "luxuriating metabolism" ("luxurierender Stoffwechsel") to indicate that plants may well be able to afford carrying out such biosyntheses even if the product does not necessarily provide a competitive advantage. However, since there is evidence to support the occurrence of metabolic regulation in alkaloid formation, one should take the position that alkaloid formation offers a selection advantage to an organism in the process of evolution. One of the more attractive hypotheses as to what this advantage might be is that proposed by Bu'Lock (1965), who states that it is not the secondary metabolites (alkaloids) in themselves that are important but rather the actual activity of secondary biosynthesis. The argument is that during the growth phase of the organism there is an absence of secondary biosynthesis not because of a lack of appropriate precursors, but because of the absence of the required enzymatic activities. When growth is halted by some particular circumstance, an imbalance results in primary biosynthesis and this may initiate the formation of enzymes of secondary biosynthesis. For example, intermediates in primary metabolism normally present at low concentrations may tend to accumulate until the metabolic pressure is relieved by the starting-up of secondary synthetic pathways through the activation or induction of special enzymes. The relief of this metabolic pressure can be visualized as a distinct advantage in the process of selection. In apparent contrast to this hypothesis is the fact that in a process like ergot alkaloid formation in *Claviceps* the organism actually synthesizes more tryptophan and has more tryptophan biosynthetic enzyme activity during the alkaloid production phase than during the growth phase. However, one has to keep in mind that the conditions of such a fermentation are probably quite different from those under which the biosynthetic pathway evolved.

A second and rather popular speculation concerning the role of alkaloids in evolution is rather more teleological in that alkaloids are formed by the plant as a protective device and this protection offers a selection advantage. While there are many arguments against the "protective device" concept such as the great number of plant species that contain no alkaloids and the fact that plants which are toxic to one species of herbivore may not be toxic to another species (James, 1950), these arguments are really all after the fact. They do not take into consideration that, at some point in time, the presence of the alkaloid may have provided a slight positive advantage for survival by protecting against some insect or animal that may since have become extinct, and that these small positive values, compounded over thousands of years of evolution, have allowed for the survival of the plant.

In this regard it is interesting to note that the secondary metabolites of plants such as alkaloids often exhibit pronounced pharmacological activity in mammals. The mechanism of action of these compounds is usually through an influence on the complex processes of the organized tissues of the animal, for example, nerve transmission in the autonomic nervous system. On the other hand, only a few microorganisms have been found which produce secondary metabolites that exhibit this type of activity in animals. Instead the secondary metabolities produced by microorganisms, antibiotics and mycotoxins, exhibit their action by affecting basic cellular physiology, for example, protein synthesis. Only a few antibiotics have been isolated from plants and only a few pharmacologically active compounds from micoorganisms. This may, of course, simply reflect what researchers have been looking for in their studies, but it could also be postulated that antibiotics allow for survival of a particular microorganism in the milieu in which it is competing with other microorganisms, and that the alkaloids would serve the same purpose for plants against herbivores and insects.

ACKNOWLEDGMENTS

Work from the authors' laboratories was supported by the National Institutes of Health through Research Career Development Award GM 42389 (to H.G.F.) and Research Grant AM 11662 and by Eli Lilly and Company.

REFERENCES

Abe, M., and S. Yamatodani. 1964. *Progr. Ind. Microbiol.* **5**:205.
Amici, A. M., A. Minghetti, T. Scott, C. Spalla, and L. Tognoli. 1967a. *Appl. Microbiol.* **15**:597.
Amici, A. M., T. Scotti, C. Spalla, and L. Tognoli. 1967b. *Appl. Microbiol.* **15**:611.
Andreoli, A. J., M. Ikeda, Y. Nishizuka, and O. Hayaishi. 1963. *Biochem. Biophys. Res. Commun.* **12**:92.
Arcamone, F., E. B. Chain, A. Ferretti, H. Minghetti, P. Pennella, and A. Tonolo. 1962. *Biochim. Biophys. Acta* **57**:174.
Baxter, C., and M. Slaytor. 1972a. *Phytochemistry* **11**:2763.

Baxter, C., and M. Slaytor. 1972b. *Phytochemistry* 11:2767.

Bu'Lock, J. D. 1965. "The Biosynthesis of Natural Products," pp. 9–11. McGraw-Hill, New York.

Bu'Lock, J. D., and J. G. Barr. 1968. *Lloydia* 31:342.

Dawson, R. F., D. R. Christman, A. D'Adamo, M. L. Solt, and A. P. Wolf. 1960. *J. Amer. Chem. Soc.* 82:2628.

deWaart, C., and W. A. Taber. 1960. *Can. J. Microbiol.* 6:675.

Digenis, G. A. 1969. *J. Pharm. Sci.* 58:39.

Eberspaecher, F., H. Uesseler, and F. Lingens. 1970. *Hoppe-Seyler's Z. Physiol. Chem.* 351:1465.

Erge, D., W. Maier, and D. Groeger. 1973. *Biochem. Physiol. Pflanzen* 164:234.

Evans, W. C., N. A. Stevenson, and R. F. Timoney. 1969. *Planta Med.* 17:120.

Fairbairn, J. W., and S. El-Masry. 1967. *Phytochemistry* 6:499.

Fairbairn, J. W., J. M. Palmer, and A. Paterson. 1968. *Phytochemistry* 7:2117.

Fleeker, J., and R. U. Byerrum. 1967. *J. Biol. Chem.* 242:3042.

Floss, H. G., and U. Mothes. 1964. *Arch. Mikrobiol.* 48:213.

Frost, G. M., K. S. Yang, and G. R. Waller. 1967. *J. Biol. Chem.* 242:887.

Gholson, R. K., I. Ueda, N. Ogasawara, and L. M. Henderson. 1964. *J. Biol. Chem.* 239:1208.

Groeger, D. 1963. *Planta Med.* 11:444.

Groeger, D., and D. Erge. 1961. *Planta Med.* 9:471.

Groeger, D., K. Mothes, H. G. Floss, and F. Weygand. 1963. *Z. Naturforsch.* B18:1123.

Gross, D., H. Lehmann, and H. R. Schütte. 1970. *Z. Pflanzenphysiol.* 63:1.

Hadwiger, L. A., S. E. Badiei, G. R. Waller, and R. K. Gholson. 1963. *Biochem. Biophys. Res. Commun.* 13:466.

Heinstein, P. F., S. L. Lee, and H. G. Floss. 1971. *Biochem. Biophys. Res. Commun.* 44:1244.

Hiles, R. A., and R. U. Byerrum. 1969. *Phytochemistry* 8:1927.

Hills, K. L., W. Bottomley, and P. I. Mortimer. 1954. *Aust. J. Appl. Sci.* 5:283.

James, W. O. 1950. *Alkaloids* 1:83–84.

Johnson, R. D., and G. R. Waller. 1972. *Abstr. Pap., 164th Meet, Amer. Chem. Soc.* Biol. 227.

Kahn, V., and J. J. Blum. 1968. *J. Biol. Chem.* 243:1448.

Kaplan, H., U. Hornemann, K. M. Kelley, and H. G. Floss. 1969. *Lloydia* 32:489.

Leete, E. 1967. *In* "Biogenesis of Natural Compounds" (P. Bernfeld, ed.), 2nd ed., pp. 953–1023. Pergamon, Oxford.

Leete, E. (1969). *Advan. Enzymol.* 32:373–422.

Leete, E., and F. H. B. Leitz. 1957. *Chem. Ind. (London)* p. 1572.

Lingens, F. 1971. *In* "International Symposium on the Biochemistry and Physiology of Alkaloids" (K. Mothes, K. Schreiber, and H. R. Schütte, eds.), 4th, pp. 65–73. Abh. Deut. Akad. Wiss., Berlin.

Lingens, F., W. Goebel, and H. Uesseler. 1967. *Eur. J. Biochem.* 2:442.

Luckner, M., and L. Nover. 1971. *In* "International Symposium on the Biochemistry and Physiology of Alkaloids" (K. Mothes, K. Schreiber, and H. R. Schuette, eds.), 4th, pp. 525–533. Abh. Deut. Akad. Wiss., Berlin.

McFarlane, I. J., E. M. Lees, and E. E. Conn. 1973. *13th Annu. Meet., PSNA, Pac. Grove* Abstract C-15.

Mann, D. F. 1972. Ph.D. Thesis, Michigan State University, East Lansing.

Mann, J. D., and S. H. Mudd. 1963. *J. Biol. Chem.* 238:381.

Mann, J. D., C. E. Steinhart, and S. H. Mudd. 1963. *J. Biol. Chem.* **238**:676.

Mizusaki, S., Y. Tanabe, M. Noguchi, and E. Tamaki. 1972. *Symp. Plant Chem., 8th, 1972* p. 7.

Mizusaki, S., Y. Tanabe, M. Noguchi, and E. Tamaki. 1973. *Plant Cell Physiol.* **14**:103.

Mockaitis, J. M., A. Kivilaan, and A. Schulze. 1973. *Biochem. Physiol. Pflanzen* **164**:248.

Mothes, K. 1969. *Experientia* **25**:225.

Mothes, K., and H. Schroeter, eds. 1963. "International Symposium on the Biochemistry and Physiology of Alkaloids," 2nd, No. 4. Abh. Deut. Akad. Wiss., Berlin.

Mothes, K., and H. R. Schütte, eds. 1956. "International Symposium on the Biochemistry and Physiology of Alkaloids," 1st, No. 7. Q Abh. Deut. Akad. Wiss., Berlin.

Mothes, K., and H. R. Schütte. 1969a. "Biosynthese der Alkaloide." VEB Deut. Verlag Wiss., Berlin.

Mothes, K., and H. R. Schütte. 1969b. "Biosynthese der Alkaloide," p. 15. VEB Deut. Verlag Wiss., Berlin.

Mothes, K., D. Gross, H. Liebisch, and H. R. Schütte, eds. 1966. "International Symposium on the Biochemistry and Physiology of Alkaloids," 3rd, No. 3. Abh. Deut. Akad. Wiss., Berlin.

Mothes, K., K. Schreiber, and H. R. Schütte, eds. 1971. "International Symposium on the Biochemistry and Physiology of Alkaloids," 4th. Abh. Deut. Akad. Wiss., Berlin.

Munsche, D. 1964. *Flora (Jena)* **154**:317.

Nakata, K., and M. Tanaka. 1968. *Jap. J. Genet.* **43**:65.

Nishizuka, Y., and O. Hayaishi. 1963. *J. Biol. Chem.* **238**:3369.

Nitsch, J. P., and C. Nitsch. 1969. *Science* **163**:85.

Řeháček, Z., J. Kozová, A. Řičicová, J. Kašlík, P. Sajdl, S. Švarc, and S. C. Basappa. 1971. *Folia Microbiol. (Prague)* **16**:35.

Robbers, J. E., and H. G. Floss. 1970. *J. Pharm. Sci.* **59**:702.

Robbers, J. E., L. G. Jones, and V. M. Krupinski. 1972a. *Lloydia* **35**:471P.

Robbers, J. E., L. W. Robertson, K. M. Hornemann, A. Jindra, and H. G. Floss. 1972b. *J. Bacteriol.* **112**:791.

Robertson, L. W. 1971. Ph.D. Thesis, Purdue University, Lafayette, Indiana.

Robertson, L. W., J. E. Robbers, and H. G. Floss. 1973. *J. Bacteriol.* **114**:208.

Romeike, A. 1961. *Kulturpflanze* **9**:171.

Romeike, A. 1962. *Kulturpflanze* **10**:140.

Romeike, A. 1966. *Kulturpflanze* **14**:129.

Rothe, U., and W. Fritsche. 1967. *Arch. Mikrobiol.* **58**:77.

Rowson, J. M. 1945a. *Quart. J. Pharm. Pharmacol.* **18**:175.

Rowson, J. M. 1945b. *Quart. J. Pharm. Pharmacol.* **18**:185.

Schiedt, U., G. Boeckh-Behrens, and A. M. Delluva. 1962. *Hoppe-Seyler's Z. Physiol. Chem.* **330**:46.

Schmauder, H. P., and D. Groeger. 1973. *Biochem. Physiol. Pflanzen* **164**:41.

Sievers, A. F. 1915. *U.S., Dep. Agr., Bull.* **306**.

Solomon, M. J., and F. A. Crane. 1970. *J. Pharm. Sci.* **59**:1670.

Solt, M. L., R. F. Dawson, and D. R. Christman. 1960. *Plant Physiol.* **35**:887.

Spalla, C. 1972. *In* "Genetics of Industrial Microorganisms" (Z. Vaněk, Z. Hošťalek, and J. Cudlin, eds.), Vol. II, pp. 393–403. Elsevier, Amsterdam.

Spalla, C., A. M. Amici, T. Scotti, and L. Tognoli. 1969. *In* "Fermentation Advances" (D. Perlman, ed.), pp. 611–628. Academic Press, New York.

Spenser, I. D. 1968. *Compr. Biochem.* **20**:231–413.

Sproessler, B., and F. Lingens. 1970. *Hoppe-Seyler's Z. Physiol. Chem.* **351**:448 and 967.

Stary, F. 1963. *In* "International Symposium on the Biochemistry and Physiology of Alkaloids" (K. Mothes and H. Schroeter, eds.), 2nd, No. 4, pp. 175–178. Abh. Deut. Akad. Wiss. Berlin.

Taber, W. A. 1964. *Appl. Microbiol.* **12**:321.

Teuscher, E. 1964. *Flora (Jena)* **155**:80.

Teuscher, E. 1965. *Pharmazie* **20**:778.

Teuscher, E. 1966. *In* "International Symposium on the Biochemistry and Physiology of Alkaloids" (K. Mothes *et al.*, eds.), 3rd, No. 3, pp. 429–438. Abh. Deut. Akad. Wiss., Berlin.

Tyler, V. E., Jr. 1961. *J. Pharm. Sci.* **50**:629.

Vazujfalvi, D. 1971. *In* "International Symposium on the Biochemistry and Physiology of Alkaloids" (K. Mothes, K. Schreiber, and H. R. Schuette, eds.), 4th, pp. 59–61. Abh. Deut. Akzd. Wiss., Berlin.

Vining, L. C. 1970. *Can. J. Microbiol.* **16**:473.

Vislin, M. L., D. Munsche, and H. B. Schroeter. 1964. *Flora (Jena)* **154**:299.

Waller, G. R., and L. M. Henderson. 1961. *J. Biol. Chem.* **236**:1186.

Waller, G. R., and K. S. Yang. 1967. *Phytochemistry* **6**:1637.

Waller, G. R., K. S. Yang, R. K. Gholson, L. A. Hadwiger, and S. Chaykin. 1966. *J. Biol. Chem.* **241**:4411.

Wanner, H., and P. Kalberer. 1966. *In* "International Symposium on the Biochemistry and Physiology of Alkaloids" (K. Mothes *et al.*, eds.), 3rd, No. 3, pp. 607–610. Abh. Deut. Akad. Wiss., Berlin.

Weygand, F., and H. G. Floss. 1963. *Angew. Chem., Int. Ed. Engl.* **2**:243.

White, H. A., and M. Spencer. 1964. *Can. J. Bot.* **42**:1481.

Wiewiórowski, M., H. Podkowińska, M. D. Bratek-Wiewiórowska, M. Kuhn-Orzechóroska, and W. Boczoń. 1966. *In* "International Symposium on the Biochemistry and Physiology of Alkaloids" (K. Mothes *et al.*, eds.), 3rd, No. 3, pp. 215–233. Abh. Deut. Akad. Wiss., Berlin.

Willuhn, G. 1966. *In* "International Symposium on the Biochemistry and Physiology of Alkaloids" (K. Mothes *et al.*, eds.), 3rd, No. 3, pp. 97–103. Abh. Deut. Akad. Wiss., Berlin. (1966).

Yang, K. S., and G. R. Waller. 1965. *Phytochemistry* **4**:881.

Yang, K. S., R. K. Gholson, and G. R. Waller. 1965. *J. Amer. Chem. Soc.* **87**:4184.

THE BIOCHEMISTRY OF *myo*-INOSITOL IN PLANTS

F. LOEWUS

Department of Biology, State University of New York at Buffalo,
Buffalo, New York

Introduction

The inositols constitute an important group of naturally occurring polyhydric alcohols and some isomers of this group are commonly found in most plants, provided adequate methods of detection are employed. Such methods, especially the use of sensitive isotopic techniques and chromatographic procedures, have provided the means of demonstrating carbohydrate interconversions in which inositol plays a central role. To examine this role, I will review certain aspects of inositol biosynthesis and metabolism as related to plants. For those wishing to obtain additional

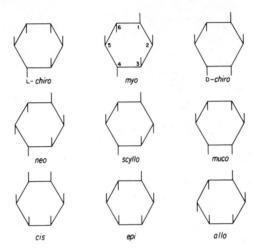

FIG. 1. Configurational forms of inositol.

information, the following books and reviews are useful (Anderson, 1966, 1972; Eisenberg, 1969; Kindl, 1966a; Kindl and Hoffmann-Ostenhof, 1966a; Loewus, 1971; Plouvier, 1963; Posternak, 1965: Tanner, 1967).

The nine stereoisomers of inositol consist of seven *meso* forms and one DL pair as shown in Figs. 1 and 2. Of these, *myo*-inositol is most widely distributed in plants and probably occurs in all living species (Anderson, 1972; Posternak, 1965). A convention for numbering each carbon atom in

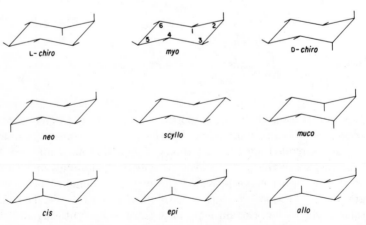

FIG. 2. Selected conformational isomers of the various configurational forms of inositol.

an inositol isomer has been established (Anderson, 1972; IUPAC Commission on the Nomenclature of Organic Chemistry and IUPAC-IUB Commission on Biochemical Nomenclature, 1968). According to this convention, *myo*-inositol may be numbered clockwise, assigning the lowest possible numbers to the sequence of carbons bearing hydroxyls which project above the plane of the ring. This convention, while convenient and consistent as a means of formally identifying specific carbons in the inositol structure, offers little comfort to the biochemist who seeks to trace out the biosynthesis and metabolism of *myo*-inositol in plants. As will be seen shortly, the numbering system used for biosynthetically derived *myo*-inositol is inverse to that of its glucose precursor.

Biosynthesis of *myo*-Inositol

Inherent in the *myo*-inositol structure, C-2 through C-5, is the D-*gluco* configuration. Although other isomers of inositol might be construed to contain the same potential D-*gluco* configuration, *myo*-inositol appears to be unique. This quality was recognized long ago by Maquenne (1900) but a biosynthetic relationship between D-glucose and *myo*-inositol was not established experimentally until quite recently. The first such experiments (Charalampous, 1957; Daughaday *et al.*, 1955; Eagle *et al.*, 1960; Halliday and Anderson, 1955) which were performed with animal tissue or yeast cells demonstrated the conversion of D-glucose to *myo*-inositol but did not reveal the mechanism. Evidence favoring a cyclization mechanism was obtained from studies in detached parsley leaves of the conversion of D-glucose-1-^{14}C to *myo*-inositol, labeled predominately in carbon-6 (Loewus, 1965; Loewus and Kelly, 1962). Soon other reports appeared, extending the observation made in plant tissue to similar processes in animal tissue (Eisenberg, 1966) and microorganisms (Kindl, 1966b). Studies of *myo*-inositol biosynthesis in cell-free preparations of rat testis (Eisenberg, 1966), yeast (Chen and Charalampous, 1966), *Neurospora* (Piña and Tatum, 1967), and higher plants (Kurasawa *et al.*, 1967; Imhoff and Bourdu, 1973; Loewus and Loewus, 1971; Matheson and St. Clair, 1971; Ruis *et al.*, 1967) established the fact that D-glucose 6-phosphate was the actual substrate. The first identifiable product, 1L-*myo*-inositol 1-phosphate, was reported initially by Chen and Charalampous (1966), Eisenberg and Bolden (1965), and Ruis *et al.* (1967). A phosphatase in crude cell-free preparations of cyclizing enzyme from rat testis and yeast hydrolyzed 1L-*myo*-inositol to free *myo*-inositol and inorganic phosphate (Chen and Charalampous, 1966; Eisenberg, 1967). These crude systems required both NAD^+ and Mg^{2+} for activity, but resolution of the cyclizing enzyme and the phosphatase revealed that only NAD$^+$ was needed by the cyclizing enzyme, while the Mg^{2+} requirement was associated with the phosphatase. Conversion of D-glucose to *myo*-inositol occurred in three enzymatic steps, as seen in Fig. 3.

The enzyme catalyzing conversion of D-glucose 6-phosphate to 1L-*myo*-inositol 1-phosphate is referred to variously as D-glucose 6-phosphate : 1L-*myo*-inositol 1-phosphate cyclase (Barnett and Corina, 1968; Sherman *et al.*, 1969) or cycloaldolase (Loewus and Loewus, 1971; Piña *et al.*, 1969). An EC number has yet to be assigned. Partial purification of cycloaldolase and studies accompanying this effort have generated a wealth of information regarding the mechanism of action, the physical and chemical properties, and the various components of this enzyme system.

For reasons still undetermined, cycloaldolase isolated from animal tissues loses activity quite rapidly during purification; therefore, most studies of cycloaldolase from such sources have been limited to relatively crude preparations (Barnett and Corina, 1968; Barnett *et al.*, 1973; Eisenberg, 1967; Sherman *et al.*, 1969, 1971). Enzyme from plant (Loewus and Loewus, 1971, 1973a) or fungal (Brunner *et al.*, 1972; Mogyoros *et al.*, 1972) sources is more stable to purification procedures, but the specific activities of such preparations are scarcely better than crude preparations from enzyme-rich tissues such as rat testis.

Although addition of NAD$^+$ to the medium was required for activity in earlier studies, it now appears that such a requirement stemmed from dissociation of tightly bound NAD$^+$ during enzyme purification. Bound NAD$^+$ has been demonstrated in cycloaldolase from rat testis, *Neurospora*, and cell suspension culture of *Acer pseudoplatanus* (Barnett and Corina, 1968; Barnett *et al.*, 1973; Brunner *et al.*, 1972; Loewus and Loewus, 1973b). The role of bound NAD$^+$ in D-glucose 6-phosphate cyclization will be discussed later.

Tracer studies have contributed mightily to current understanding of the stereochemistry of the conversion of D-glucose to *myo*-inositol. A summary of information obtained from these experiments is presented in Fig. 4.

D-Glucose labeled with ^{14}C in position C-1 ($\triangle$), C-2($\square$), or C-6 ($\bigcirc$) was

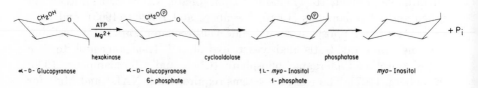

FIG. 3. Enzymatic conversion of D-glucose to *myo*-inositol.

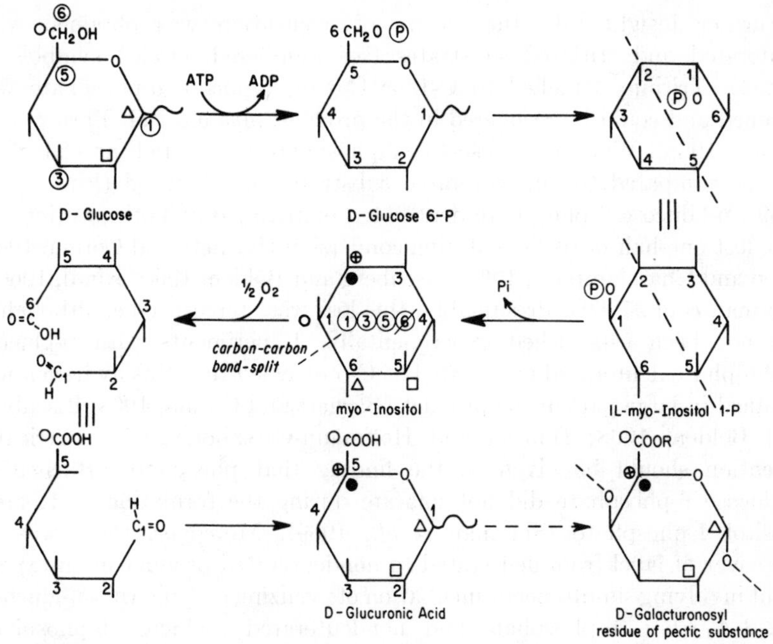

FIG. 4. Conversion of D-glucose to *myo*-inositol and subsequent oxidative cleavage of *myo*-inositol to form D-glucuronic acid and D-galacturonosyl units of pectin. Specifically labeled positions are described in the text. This figure is reproduced from F. Loewus, ed., "Biogenesis of Plant Cell Wall Polysaccharides." Academic Press, New York, 1973.

converted to *myo*-inositol in which ^{14}C was located mainly in C-6 ($\triangle$), C-5 ($\square$), or C-1 ($\bigcirc$), respectively (Chen and Charalampous, 1964; Eisenberg *et al.*, 1964; Hauska and Hoffmann-Ostenhof, 1967; Kindl and Hoffmann-Ostenhof, 1964a, b; Loewus and Kelly, 1962). As mentioned earlier, vicissitudes of cyclitol nomenclature have led to ring numbering of carbons in *myo*-inositol which is inversely related to the assignment of numbers in the carbon chain of D-glucose. Location of ^{14}C in *myo*-inositol was determined by a combination of biochemical and chemical procedures, enzymatic conversion of *myo*-inositol to either D-galacturonic acid (Eisenberg *et al.*, 1964; Loewus and Kelly, 1962) or D-glucuronic acid (Chen and Charalampous, 1964), or by straight chemical methods (Hauska and Hoffmann-Ostenhof, 1967; Kindl and Hoffmann-Ostenhof, 1964a,b). Evidence for the specific cleavage of *myo*-inositol, obtained by use of *myo*-inositol-2-^{14}C ($\bullet$) and *myo*-inositol-2-^{3}H ($\otimes$) which were converted to D-glucuronic acid-5-^{14}C ($\bullet$) and D-galacturonic acid-5-^{14}C ($\bullet$) or -5-^{3}H ($\otimes$) by kidney tissue (Charalampous, 1960) and higher plants (Loewus *et al.*, 1962), provided the basis for these experiments.

Further insight into the nature of cyclization was obtained with deuterated and tritiated substrates (see numbered, circled symbols in Fig. 4). Tritium attached to C-1 or C-3 of D-glucose 6-phosphate was retained and could be recovered in the product, *myo*-inositol (Barnett and Corina, 1968). When D-glucose-1-d_1 6-phosphate was tested, no change in rate as compared to nondeuterated substrate was observed (Piña *et al.*, 1969). D-Glucose 6-phosphate deuterated or tritiated at both positions on C-6, lost one-half of its label during conversion (Barnett and Corina, 1968; Chen and Charalampous, 1967; Eisenberg and Bolden, 1968; Kindl, 1966b; Sherman *et al.*, 1969). Presumably, this loss was stereospecific, although it has not been established experimentally. Experiments with D-glucose 6-phosphate deuterated or tritiated at C-5 revealed that this hydrogen was retained in large part in the product (Barnett and Corina, 1968; Eisenberg and Bolden, 1968; Hauska and Hoffmann-Ostenhof, 1967). Particular attention should be given to the finding that phosphate attached to D-glucose 6-phosphate did not migrate during the formation of 1L-*myo*-inositol 1-phosphate (Sherman *et al.*, 1969). Moreover, there was no crossover of label from deuterated to nondeuterated product in an experiment involving simultaneous incubation of cyclizing enzyme with D-glucose-1,2,3,4,5,6-d_7 6-phosphate and nondeuterated D-glucose 6-phosphate (Sherman *et al.*, 1969). This latter finding provided evidence that all intermediate products of the reaction were enzyme-bound.

A mechanism of cyclization of D-glucose 6-phosphate to 1L-*myo*-inositol 1-phosphate has emerged from the studies just described and its principal features are shown in Fig. 5. Briefly, an NAD⁺-catalyzed oxidation of D-glucose 6-phosphate produces D-*xylo*-hexos-5-ulose 6-phosphate (often

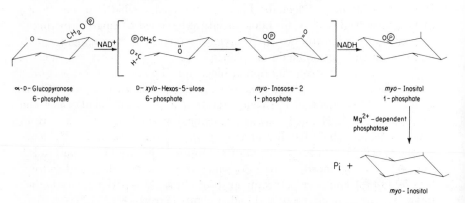

FIG. 5. Proposed mechanism for the conversion of D-glucose 6-phosphate to *myo*-inositol 1-phosphate by cycloaldolase.

called 5-keto-D-glucose 6-phosphate) which undergoes an aldol condensation between C-1 and C-6 to form *myo*-inosose-2 1-phosphate. Stereospecific reduction of the latter by NADH results in 1L-*myo*-inositol 1-phosphate which is released from the enzyme as a final product. Intermediate products, D-*xylo*-hexos-5-ulose 6-phosphate and *myo*-inosose-2 1-phosphate, as well as the reduced form of NAD are tightly bound to the enzyme. As noted earlier, NAD⁺ is also bound to the enzyme in the native state. The experiments with D-glucose 6-phosphate containing deuterium or tritium at position 5 as well as experiments with perdeuterated substrate clearly showed that hydrogen removed from C-5 in the initial step was replaced in the final reductive step (Barnett and Corina, 1968; Hauska and Hoffmann-Ostenhof, 1967; Sherman *et al.*, 1969). Attempts to introduce tritium into position 5 of the substrate, D-glucose 6-phosphate, or position 2 of the product, *myo*-inositol 1-phosphate, by adding R-NADH(³H), S-NADH(³H), or a mixture of R- (62.5%) and S-NADH(³H) (37.5%) to the incubation medium of cycloaldolase and D-glucose 6-phosphate were unsuccessful (Barnett and Corina, 1968; Chen and Charalampous, 1967). These earlier experiments lent strength to the view that NAD⁺ became tightly bound to the enzyme after reduction and could not be replaced by added NADH. When the cyclization was performed in tritiated water, uptake of one tritium atom per *myo*-inositol was observed in *Candida utilis* preparations, but none of this label appeared in position 2 of the product (Kindl, 1966b). A similar experiment using rat testis enzyme failed to incorporate tritium into product (Barnett and Corina, 1968).

Treatment with charcoal removed NAD⁺ from the rat testis enzyme, leaving a totally inactive preparation (Barnett and Corina, 1968; Barnett *et al.*, 1973). Combination of this "inactive" preparation with substrate in the presence of added NAD⁺ restored 80 percent of the original activity. Barnett *et al.* (1973) have used this observation to explore partial reactions of cycloaldolase as outlined in Fig. 5. "Inactive" enzyme catalyzed the reduction of D-*xylo*-hexos-5-ulose 6-phosphate by NADH(³H) (the latter prepared by reduction of NAD⁺ with dithionite in tritiated water) to form D-glucose-5-³H 6-phosphate in which the location of tritium was established by chemical degradation. It also catalyzed the reduction of L-sorbose 6-phosphate (also referred to as 5-keto-D-glucitol 6-phosphate) by NADH(³H) to form D-glucitol-5-³H 6-pholphate. The latter compound was a good inhibitor of cycloaldolase. 2-Deoxy-D-glucose 6-phosphate, a powerful competitive inhibitor of the overall reaction (Barnett *et al.*, 1970), inhibited partial reactions involving both D-*xylo*-hexos-5-ulose 6-phosphate and L-sorbose 6-phosphate. Tracer levels of tritiated *myo*-inositol 1-phosphate could also be detected when D-*xylo*-hexos-5-ulose was incubated for 3 hours with "inactive" cycloaldolase in the presence of NADH(³H).

Although less than 0.01 percent of the tritium present appeared in the product, this amount was significantly greater than boiled or denatured control samples. No attempt was made to determine the position of tritium in this product. The possibility that appearance of label in the *myo*-inositol 1-phosphate represented an overall reaction involving labeled D-glucose 6-phosphate generated by reduction of D-*xylo*-hexos-5-ulose 6-phosphate must also be considered (F. Eisenberg, Jr., personal communication).

Some information regarding the aldolase-like partial reaction in which D-*xylo*-hexos-5-ulose 6-phosphate is cyclized to *myo*-inosose-2 1-phosphate is now available. Attempts to demonstrate the existence of a lysine-bound Schiff base intermediate between enzyme and substrate such as characterizes Class I aldolases (Horecker *et al.*, 1972; Lai and Horecker, 1972) have been unsuccessful in cycloaldolase preparations from rat testis (Barnett *et al.*, 1973), *Neurospora* (Mogyoros *et al.*, 1972) and suspension-cultured cells of *Acer pseudoplatanus* (Loewus and Loewus, 1973a). On the other hand, there are indications that cycloaldolases from the latter two sources have characteristics similar to Class II aldolases in which a metal ion at the active center performs as an equivalent to the Schiff base intermediate. Both *Neurospora* and *Acer* enzymes were strongly inhibited by EDTA and exhibited other properties normally associated with Class II aldolases.

Chemical evidence to support the partial reactions presented in Fig. 5 has been obtained by Kiely and Fletcher (1969). Treatment of D-*xylo*-hexos-5-ulose with dilute alkali produced *myo*-inosose-2, which upon reduction with sodium borohydride yielded a mixture of *scyllo*- and *myo*-inositol. Recently, Sherman *et al.* (1968) were able to demonstrate the presence of free *myo*-inosose-2 in nature. The presence of *scyllo*-inositol in plant and animal tissues is well established. Given the reaction sequence in Fig. 5, it is not surprising to encounter *myo*-inosose-2 and its reduction products in nature, possibly the result of nonspecific phosphatase activity.

The reaction catalyzed by D-glucose 6-phosphate : 1L-*myo*-inositol 1-phosphate cycloaldolase appears to be the only biosynthetic process of *myo*-inositol formation in plants, possibly the only process in all organisms which produce this cyclitol. Once formed, the plant has many ways of storing and utilizing this compound. These processes will be examined in the following sections.

Biosynthesis and Breakdown of Phytic Acid

Phytic acid, the hexakis-orthophosphate ester of *myo*-inositol, occurs widely in higher plants and is regarded as the main storage form of phosphate in mature and dormant seeds. Substantial evidence now supports the Anderson structure given in the upper part of Fig. 6 (Angyal and Russell, 1969; Blank *et al.*, 1971; Johnson and Tate, 1969a). Evidence for a conformation in which phosphate groups at positions 1,3,4,5,6 are oriented axially to the plane of the ring comes from X-ray analysis of the crystalline dodecasodium salt (Blank *et al.*, 1971), but this observation may only apply in solution in a limited pH range (Brown and Tate, 1973).

Until recently, attempts to trace the biosynthesis of phytic acid have been limited largely to studies of intact plant tissues. This literature has been reviewed (Cosgrove, 1966a; Loewus, 1971). Briefly stated, experiments suggested a stepwise path of biosynthesis from *myo*-inositol or a *myo*-inositol monophosphate to phytic acid, in which intermediate di-, tri-, tetra-, and pentaphosphate esters appear in trace amounts. An intermediate common to both *myo*-inositol and D-glucose 6-phosphate was indicated and results pointed to a *myo*-inositol 1-phosphate as the likely compound (Kindl and Hoffmann-Ostenhof, 1966b; Loewus, 1969; Mandel and Biswas, 1970). At least two biosynthetic routes to *myo*-inositol 1-phosphate have been found in plants: one, the cycloaldolase conversion described in the preceding section, and the other, a *myo*-inositol kinase (Hoffmann-Ostenhof *et al.*, 1958; Dietz and Albersheim, 1965). Other sources of *myo*-inositol monophosphate such as those produced by enzymatic hydrolysis may also be involved, although this possibility seems less likely (Lim and Tate, 1973).

An important contribution to a fuller understanding of phytic acid biosynthesis is found in the work of Majumder *et al.* (1972). Following an

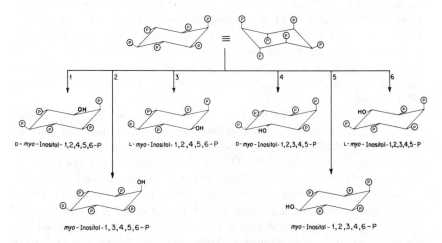

FIG. 6. Stereospecific modes of hydrolysis of phytic acid to *myo*-inositol pentaphosphates. Phytic acid is shown in both conformational forms. See text for additional details.

earlier lead uncovered in ^{32}P-incorporation studies into germinating mung bean seeds (Mandel and Biswas, 1970), they found an enzyme capable of phosphorylating *myo*-inositol phosphates ranging from the monophosphate to the pentaphosphate. The partially purified enzyme catalyzed phosphate transfer from ATP, and at lower rates from UTP and phosphoenolpyruvate, in the presence of Mg^{2+} or Mn^{2+} to *myo*-inositol mono-, di, tri-, tetra-, and pentaphosphate, converting the substrate to its next higher homolog. Since substrates were prepared chemically and then fractionated according to Cosgrove (1966b), the isomeric composition of each substrate mixture remained unresolved and little can be said at this time in regard to the stereochemistry of the biosynthetic process. The enzyme has been called a phosphoinositol kinase. It has a pH optimum at 7.4. Contaminating *myo*-inositol kinase and GTP phosphotransferase were separated from phosphoinositol kinase activity by gel electrophoresis. The enzyme was not active with *myo*-inositol as a substrate. With each substrate, it was found that the next higher homolog acted as an inhibitor, while still higher homologs showed lower inhibitory activity or no inhibition if containing more than two additional phosphate groups. Assuming a mechanism exists for removal of phytic acid from the enzyme after it is formed *in vivo*, the reaction kinetics of this enzyme suggested that accumulation of inositol phosphates other than phytic acid would not be seen in significant amounts during phytic acid biosynthesis (Majumder and Biswas, 1973a). The same paper offered preliminary evidence that phosphoinositol kinase is a phosphoprotein.

In the course of their work on phosphoinositol kinase, Majumder and Biswas (1973b) uncovered a protein inhibitor of phosphoinositol kinase in ungerminated mung bean seeds. One molecule of inhibitor appeared to inhibit one molecule of enzyme. The inhibitor was noncompetitive had a molecular weight of 86,000, and exhibited a K_i of 1.47×10^{-6} *M*. The authors suggest that this inhibitor accumulated during final stages of ripening and was destroyed or rendered ineffective during the first 24 hours after germination.

Stepwise biosynthesis of phytic acid from *myo*-inositol monophosphate probably takes place in highly localized regions of the plant cell since the occurrence of phytic acid is restricted to specific areas of the protein bodies (aleurone grains) of plant cells. During seed germination, these organelles slowly disintegrate, releasing their contents to the developing embryo (Jones, 1969). Matheson and Strother (1969) have followed the fate of phytic acid in germinating wheat over a period of 14 days during which time all phytic acid disappeared, yet the event was not accompanied by a net increase in free *myo*-inositol. Apparently the latter was utilized in carbohydrate interconversions.

Phytase, the enzyme responsible for enzymatic hydrolysis of phytic acid, catalyzes stepwise removal of phosphate groups from the hexaphosphate until free *myo*-inositol is formed (Cosgrove, 1966a). The stereospecific nature of attack by phytase on phytic acid has been under study by a small group of investigators, beginning with the work of Tomlinson and Ballou (1962). Results of these studies reveal a highly stereospecific pattern of phosphate removal, most evident at the first step of hydrolysis in which phytic acid is converted to *myo*-inositol pentaphosphate. There are two *meso* forms and two DL forms of *myo*-inositol pentaphosphate. All six are seen in Fig. 6. The two *meso* forms are substituted at positions 1, 3, 4, 5, 6 and 1, 2, 3, 4, 6 while the two DL pairs have substitutents at 1, 2, 4, 5, 6 and 1, 2, 3, 4, 5.

Each phytase that has been examined exhibits stereoselective properties and the results of these studies is summarized in Table 1.

Removal of another phosphate from either L-1, 2, 3, 4, 5 or D-1, 2, 4, 5, 6 pentaphosphate by F_1 enzyme from wheat germ, or phytases of *A. ficuum* and *N. crassa* results in D-1, 2, 5, 6-tetraphosphate which is subsequently hydrolyzed to D-1, 2, 6-triphosphate (and D-1, 2, 5-triphosphate in the case of *N. crassa*), D-1, 2-diphosphate, 2-phosphate, and ultimately, free *myo*-inositol.

Removal of another phosphate from either 1, 2, 3, 4, 6- or L-1, 2, 3, 4, 5-pentaphosphate by F_2 enzyme from wheat germ yields L-1, 2, 3, 4-tetra-

TABLE 1

Hydrolysis of Phytic Acid

Reaction step	Pentaphosphate produced	Phytase	Reference
1	D-1,2,4,5,6	*Aspergillus ficuum*[a] peaks A and E	Irving and Cosgrove (1972)
		Pseudomonas sp.[a]	Cosgrove (1970)
		Neurospora crassa[a]	Johnson and Tate (1969b)
2	1,3,4,5,6[b]	Wheat bran (F_2 fract.)	Lim and Tate (1971, 1973)
3	L-1,2,4,5,6	Wheat bran (F_1 fract.)	D. J. Cosgrove (personal communication)
4	D-1,2,3,4,5	*Pseudomonas* sp.	Cosgrove (1970)
		Aspergillus ficuum peak E	Irving and Cosgrove (1972)
5	1,2,3,4,6	Wheat bran (F_2 fract.)	Lim and Tate (1973)
6	L-1,2,3,4,5	Wheat bran (F_1 fract.)[a]	Tomlinson and Ballou (1962)
		Wheat bran (F_2 fract.)[a]	Lim and Tate (1973)

[a] Major pentaphosphate formed.
[b] Also found in chicken blood (Johnson and Tate, 1969a).

phosphate which is subsequently hydrolyzed to 1, 2, 3-triphosphate, a mixture of D-1, 2- and L-1, 2-diphosphate, 2-phosphate and, ultimately, free *myo*-inositol.

The question as to whether phosphate bond energy stored in phytic acid is conserved after removal from phytic acid or lost in the hydrolysis has not been resolved. Current studies on the conformation of phytic acid may shed fresh light on this interesting matter.

At this stage of research, phytic acid formation and breakdown can be summarized as follows:

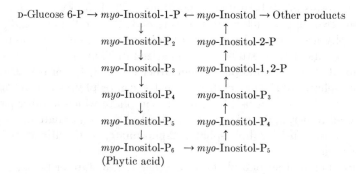

In addition to possible regulation of phytic acid biosynthesis through processes of inhibition of phosphoinositol kinase by the protein fraction from mung bean seeds, ADP inhibition (Majumder and Biswas, 1973a) also occurs *in vitro*. Breakdown of phytic acid is a carefully regulated process in plants involving regulation of phytase synthesis as well as phytase activity on phytic acid (Loewus, 1971). Each of these processes probably has a role in the overall regulation of *myo*-inositol metabolism and all are ultimately dependent upon biosynthetic processes of *myo*-inositol formation.

Isomers and Methyl Ethers of *myo*-Inositol

At this time there is no evidence of a biosynthetic pathway in plants from hexose phosphate to isomers of inositol that does not pass through *myo*-inositol. One exception that occurs in mammalian tissue has been carefully examined by Sherman *et al.* (1971). It involved conversion of D-mannose 6-phosphate to L-*neo*-inositol 1-phosphate that was catalyzed by a cycloaldolase preparation from bovine testis. The experiment was prompted by the observation that *neo*-inositol occurred naturally in calf brain as well as other organs. When a similar experiment was performed using cycloaldolase from a plant source (*Acer pseudoplatanus* cell culture) and D-mannose 6-phosphate as substrate, within the limits of the gas–liquid chromatographic system employed, no *neo*-inositol could be detected (M. W. Loewus, unpublished observation).

scyllo-Inositol, the only isomer in which one conformation places all hydroxyl groups in the equatorial position, accumulates in both plant and animal tissues (Malangeau, 1958; Plouvier, 1963). Once formed, the free compound appears to be metabolically inert in plants, although its presence as a part of more complex structures produced by microorganisms is well established (Anderson, 1972). Discovery of traces of *myo*-inosose-2 in mammalian tissues (Sherman *et al.*, 1968) prompts the suggestion that the occurrence of *scyllo*-inositol in plant as well as animal tissues may be an accidental consequence of dissociation of bound *myo*-inosose-2 1-phosphate from the cycloaldolase complex followed by enzymatic dephosphorylation. *myo*-Inosose-2 would then undergo nonspecific reduction with any of a number of secondary alcohol dehydrogenases which are present. *myo*-Inositol formed in the latter step would quickly reenter normal processes of carbohydrate metabolism, leaving the *scyllo*-inositol to accumulate.

Several years ago, Scholda *et al.* (1964a) described *in vivo* conversion of *myo*-inositol to L-*chiro*-inositol and L-quebrachitol in *Artemisia vulgaris*. Discovery of a different path of biosynthesis from *myo*-inositol to L-quebrachitol in *Acer pseudoplatanus* and *Myosotis arvensis* prompted Schilling *et al.* (1972) to repeat the *A. vulgaris* experiment and they subsequently confirmed Scholda's observation. In *A. pseudoplatanus*, as well as *M. arvensis* (Kindl and Hoffmann-Ostenhof, 1966d), *myo*-inositol is first methylated at C-3 to form D-bornesitol, then epimerized at the carbon corresponding to C-1 of *myo*-inositol to form L-quebrachitol. By contrast, in *A. vulgaris*, epimerization at C-1 precedes methylation at C-3 of *myo*-inositol. L-Quebrachitol accumulates in significant amount in sap of wintering maple trees (*Acer saccharum*). During the break in dormancy there is a sudden drop in the L-quebrachitol level (Stinson *et al.*, 1967), prompting the suggestion that L-quebrachitol may function as a reserve of preformed inositol as well as methyl groups for new tissue formation.

D-Bornesitol also acts as an intermediate for dambonitol formation in *Nerium oleander*, undergoing methylation at C-1 as well as C-3 (Kindl and Hoffmann-Ostenhof, 1966c). More recent work on cell-free preparations has shown that S-adenosylmethionine acts as methyl donor to *myo*-inositol, forming D-bornesitol in preparations from pea seedlings (Wagner *et al.*, 1969) and L-bornesitol in preparations from *Vinca rosea* (Hoffmann *et al.*, 1969).

myo-Inositol is also methylated at C-5 to form sequoyitol (Scholda *et al.*, 1964b, c, d). Epimerization at C-3 then leads to D-pinitol. Using 5-*O*-methyl-*myo*-inosose-3 as a substrate for cell-free preparations from *Trifolium incarnatum,* Kremlicka and Hoffmann-Ostenhof (1966) were able to demonstrate sequoyitol synthesis when the reductant was NADH and D-pinitol synthesis when it was NADPH. More recently, Dittrich *et al.* (1972) have shown that 1-*O*-methyl-*muco*-inositol occurs in many plants, especially gymnosperms. Chase experiments with $^{14}CO_2$ and the use of labeled precursors showed that D-pinitol was epimerized at the carbon corresponding to C-4 of *myo*-inositol to form 1-*O*-methyl-*muco*-inositol (Dittrich and Kandler, 1972). Whether any or all of these secondary products of *myo*-inositol metabolism are recalled by the plant for further use in carbohydrate metabolism poses an interesting question. It is known that substantial quantities of one or more of these methylated inositols accumulate in some plant tissues.

When *myo*-inositol in methylated at position 6, the product is D-ononitol. Kindl and Hoffmann-Ostenhof (1966b) demonstrated this conversion *in vivo* in *Ononis spinosa.* The corresponding L-isomer has not been found in plants.

The various interconversions of *myo*-inositol and its isomers and methyl ethers are assembled in Fig. 7. A quick comparison with Fig. 1 reveals that five of the nine possible isomers are readily formed in plants. There is little that can be said at this time regarding possible metabolic roles for these

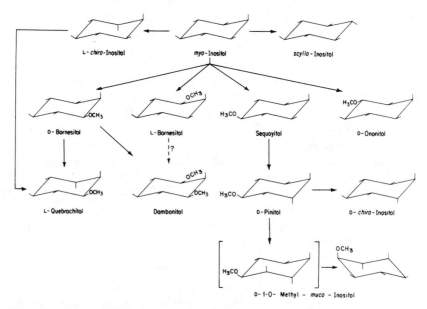

Fɪɢ. 7. Conversion of *myo*-inositol to isomeric forms and methylated derivatives that are commonly found in plant tissues.

isomers and their methyl ethers. The possibility that certain ones, notably, L-bornesitol, D-bornesitol, and sequoyitol, may act as methylated precursors of uronic acid or pentose residues in cell wall polysaccharides which contain O-methyl ethers or esters (Loewus, 1964a) has not been successfully demonstrated by experiment, although attempts have been made (Loewus, 1971).

Indol-3-acetyl Esters of *myo*-Inositol

The indol-3-acetyl esters of *myo*-inositol which were discovered and characterized by Bandurski and his colleagues (Bandurski and Piskornik, 1973) represent a fascinating group of compounds and a real challenge to both the plant biochemist and the plant physiologist. There is still no information available regarding the biosynthesis of these esters, nor is there a great deal to be said regarding their physiological role in plants.

The basic structure of the so-called B_2 ester (Labarca *et al.*, 1965; Ueda *et al.*, 1970; Ehmann and Bandurski, 1972), clearly established by isolation and NMR analysis, is 2-O-(indol-3-acetyl)-*myo*-inositol (Nicholls, 1967; Nicholls *et al.*, 1971). The formula of B_2 is given in Fig. 8. Two additional

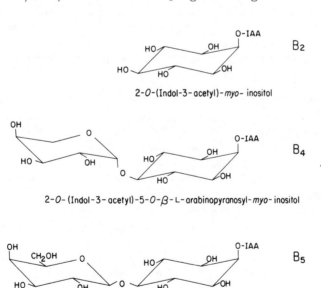

2-O-(Indol-3-acetyl)-*myo*-inositol

2-O-(Indol-3-acetyl)-5-O-β-L-arabinopyranosyl-*myo*-inositol

2-O-(Indol-3-acetyl)-5-O-β-D-galactopyranosyl-*myo*-inositol

FIG. 8. Representative indol-3-acetic acid esters of *myo*-inositol and *myo*-inositol glycosides found in the B fraction of corn kernels. After Bandurski and Piskornik (1973).

myo-inositol esters of indol-3-acetic acid have been isolated which differ from B₂ in the position occupied by the ester bond. Whether these are naturally produced substances or artifacts of isolation has not been clearly established. Two other sets of esters have been identified; both are 5-O-glycosyl-*myo*-inositol esters in which the sugar residue is arabinose or galactose (B₄ and B₅ in Fig. 8). Since each of these glycosyl derivatives also exists in the three ester configurations, a total of nine indol-3-acetyl-*myo*-inositol compounds are known.

Bandurski has focused his attention on the bound indol-3-acetic acid-containing substances of corn kernels in an effort to characterize all such compounds from this source. The B fraction, represented by *myo*-inositol esters, accounts for about 50 percent of these compounds. The remaining portion, fraction A, consists of indol-3-acetyl esters of cellulosic glucan and other unidentified compounds in approximately equal amount (Piskornik and Bandurski, 1972). Evidence to show that the ester link occurs at C-2, C-4, and C-6 of D-glucose has been reported recently (Ehmann and Bandurski, 1973).

The intimate chemical relationship between indol-3-acetic acid, *myo*-inositol, and sugars common to cell wall polysaccharides provides an interesting challenge and problem to those working in the area of carbohydrate metabolism, a matter certain to heighten speculation regarding the connection between hormonal regulation of growth and changes in the chemical and and physical properties of the primary cell wall.

myo-Inositol as a Cofactor in Galactosyl Transfer Reactions

When galactinol (1L-O-α-D-galactopyranosyl-*myo*-inositol) was first isolated from sugar beet molases in 1953 by Brown and Serro, its role as a galactosyl donor in certain plants remained obscure. This unusual role was delineated by Tanner and Kandler in 1968. Galactinol is formed from UDP-D-galactose and *myo*-inositol as seen in Fig. 9 (Frydmann and Neufeld, 1963). Significant amounts of this compound are found in plants which accumulate sugars of the raffinose series (Sensor and Kandler, 1967; Imhoff, 1970; Webb, 1973). Galactinol participates in a series of galactosyl transfer reactions, successively transferring single galactosyl residues to sucrose, raffinose, and higher homologs of raffinose. At each step, free *myo*-inositol is produced as one of the products. The reaction is reversible. In the case of stachyose biosynthesis in *Phaseolus vulgaris*, the K_{equil} for [stachyose][*myo*-inositol]/[raffinose][galactinol] is approximately 4 (Tanner and Kandler, 1968). This reversibility provides a route for rapid exchange of labeled moieties, a phenomenon first noted by Moreno and Cardini

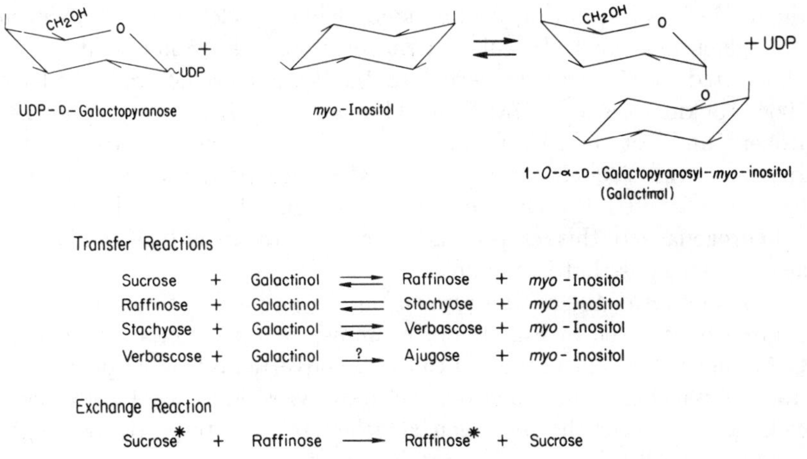

FIG. 9. The role of *myo*-inositol in the biosynthesis of sugars in the raffinose family of oligosaccharides.

(1966). In addition to the exchange reaction between labeled sucrose and raffinose shown in Fig. 9, other exchanges between galactinol and *myo*-inositol and between stachyose and raffinose have been reported (Tanner and Kandler, 1968).

More recently, Imhoff and Bourdu (1970), while studying the concentration of nonphosphorylated soluble carbohydrates and inositol in chloroplasts of *Lolium italicum*, noted that strong illumination followed by a long period of darkness trapped most of the *myo*-inositol, as well as most of the sucrose of the leaf, in the chloroplast. Their observation suggested that biosynthesis of *myo*-inositol is localized in the plastid. This view was strengthened by their demonstration of *myo-inositol* biosynthesis by cycloaldolase isolated from chloroplast extracts of pea leaves (Imhoff and Bourdu, 1973). Their finding lends weight to the view that *myo*-inositol, synthesized in the chloroplast, is required by the plant cell as a cofactor for transferring sugar residues through the chloroplast membrane into the cytoplasm. Other physiological aspects of *myo*-inositol and its interaction with carbohydrate interconversions of this type were reviewed earlier (Loewus, 1971; Loewus *et al.*, 1973).

myo-Inositol as a Precursor of Uronic Acid and Pentose for Cell Wall Polysaccharide Biosynthesis

In the conversion of *myo*-inositol to UDP-D-glucuronic acid and products of glucuronic acid metabolism, the characteristic cyclohexane structure of the inositols is lost during the first step, oxidative cleavage of *myo*-inositol to D-glucuronic acid. In subsequent reactions, glucuronic acid is phosphorylated at C-1 and converted to UDP-D-glucuronic acid (Anderson, 1966; Dickinson *et al.*, 1973; Feingold and Fan, 1973; Loewus *et al.*, 1973; Roberts and Cetorelli, 1973). The enzymatic sequence is shown in Fig. 10. In effect, this sequence provides an alternate pathway from D-glucose 6-phosphate to UDP-D-glucuronic acid, one which bypasses UDP-D-glucose dehydrogenase. In this respect, plants have a pathway of UDP-D-glucuronic acid biosynthesis that is unique.

Some details of the overall conversion of *myo*-inositol to uronic acid and pentose were given in Fig. 4. For example, *myo*-inositol labeled either in C-2 or in the hydrogen attached to C-2 is converted by higher plants to cell wall polysaccharide in which D-galacturonosyl residues are labeled specifically in C-5 or in the hydrogen attached to C-5, respectively. Pentose residues of wall polysaccharides retain both labels on the terminal hydroxymethyl group, C-5 (Loewus, 1965). In the case of *myo*-inositol-2-³H precursor, it has been shown that the pathway to pentose is stereospecific with regard to location of tritium at position 5, that is, label appears exclusively in the R chiral position of C-5 (Feingold and Fan, 1973; Loewus, 1964b).

An important consequence of this path of conversion should not be overlooked. In the conversion of D-glucose 6-phosphate to UDP-D-glucuronic acid, whether by the sugar nucleotide oxidation pathway or by the *myo*-inositol oxidation pathway, the D-*gluco* configuration is conserved by the final product. Tracer studies involving simple experiments with labeled D-glucose offer no clues as to which pathway is functional (Loewus and Kelly, 1962). Until the *myo*-inositol oxidation pathway was first proposed, it was generally assumed that all uronic acid products were formed via UDP-D-glucose. When it became evident that the hexose phosphate pool also provided the six-carbon precursor of *myo*-inositol and that *myo*-inositol

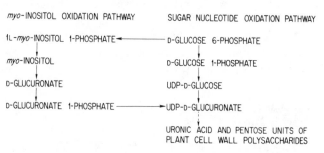

FIG. 10. Alternate pathways from D-glucose 6-phosphate to UDP-D-glucuronic acid.

TABLE 2

Uptake and Incorporation of Label from *myo*-
Inositol-2-^{14}C by Corn Root Tips from
3-day-old Seedlings[a]

myo-Inositol conc.		Uptake in 6 hours	Appearance of ^{14}C in 80 percent ethanol-insoluble residues
(*M*)	(μg/3 ml)	(μg)	(μg)
10^{-1}	54,000	1620	230
10^{-3}	540	85	29
10^{-5}	5.4	1.1	0.37

[a] Endogenous free *myo*-inositol content per 25 root tips = 14.4 μg.

was an effective precursor of uronic acid and pentose constituents, the need for information regarding the functional nature of the *myo*-inositol oxidation pathway and its role relative to UDP-D-glucose oxidation became most important.

Our efforts to provide fresh insight into the role of *myo*-inositol as an intermediate in carbohydrate metabolism and to determine the nature of that role during cell growth and development have taken many paths, two of which have reached the point where sound experimental results are available.

The first approach, initiated by R. M. Roberts (Roberts and Loewus, 1973), sought to examine the effect of raising the internal concentration of *myo*-inositol on the flow of carbon from labeled D-glucose to cell wall polysaccharides within corn root tips, specifically β-glucan and pectic substance. The argument was made that on the one hand any label from D-glucose entering the *myo*-inositol pool of the root cell would be diluted by endogenous inositol as it passed through the *myo*-inositol oxidation pathway to UDP-D-glucuronic acid. On the other hand, UDP-D-glucuronic acid synthesis not involving intermediate formation of *myo*-inositol would be unaffected by the presence of elevated levels of *myo*-inositol.

To test this argument (Table 2) it was established that corn root tips from 3-day-old sterile seedlings contained about 14.4 μg of *myo*-inositol per 25 root tips. Batches of 25 tips were suspended in 3 ml of *myo*-inositol-2-^{14}C at three concentrations; 10^{-1}, 10^{-3}, and 10^{-5} *M*. During the next 6 hours, uptake ranged from 4.5 percent of the endogenous *myo*-inositol at the highest level to 30 percent at the lowest. Thus at the lowest level, uptake was only 8 percent of the endogenous level, while at the highest level it was more than two magnitudes greater. Conversion of labeled *myo*-inositol to cell wall polysaccharide, as measured by appearance of label in 80 percent ethanol-insoluble residues, ranged from 14 to 34 percent of the incorporated ^{14}C. Radiolabel remaining in the 80 percent ethanol-soluble fraction was largely *myo*-inositol. These data provided the basis for a subsequent experiment in which root tips were suspended in unlabeled *myo*-inositol at the same concentrations used before but with a trace amount of D-glucose-6-^{14}C added. As seen in Table 3, *myo*-inositol had virtually no effect on $^{14}CO_2$ release or on incorporation of D-glucose-6-^{14}C into 80 percent ethanol-insoluble carbohydrate. Hydrolysis of this residue with pectinase released D-galacturonic acid and neutral sugars, including D-glucose. More D-glucose was recovered from the pectinase-resistant α-cellulose fraction by acid hydrolysis. Although the specific radioactivity of D-glucosyl units, whether from pectinase hydrolysis or acid hydrolysis of pectinase-resistant glucan, remained essentially unchanged as the *myo*-inositol concentration was raised from 10^{-5} to 10^{-1} *M*, that of D-galacturonic acid fell almost one-half from its highest value.

Results are consistent with the view that the flow of carbohydrate from stored reserves or an exogenous source to UDP-D-glucuronic acid proceeds via *myo*-inositol and D-glucuronic acid rather than through UDP-D-glucose in corn seedlings. Moore (1973) has noted the promotive effect of *myo*-inositol on accumulation of a pectinlike extracellular complex from soybean suspension cultures.

In a second approach to the demonstration of the functional nature of the *myo*-inositol oxidation pathway, Chen (1972) made use of germinating pollen of *Lilium longiflorum* to study the potential role of *myo*-inositol. This

TABLE 3

Utilization of Label from D-Glucose-6-^{14}C in the Presence of *myo*-Inositol

| myo-Inositol concn.[a] (M) | Radioactivity | | Specific radioactivity (cpm/μm) | | |
| | Resp. CO$_2$ (%) | Wall (%) | Pectinase hydrol. | | Acid hydrol. |
			GalUA	Glucose	Glucose
10^{-1}	6.4	23.4	905	8100	5900
10^{-3}	7.1	25.6	1380	9000	5600
10^{-5}	8.1	28.2	1740	9300	5600

[a] Medium also contained 10^6 cpm of D-glucose-6-^{14}C in 0.2 μmole.

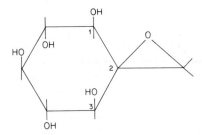

FIG. 11. Structure of 2-O,C,methylene-*myo*-inositol (MMO).

pollen has practical advantages for such an investigation. It is readily obtained, easily stored, simple to handle, and highly adaptable to experiments of the type described here. Germination and pollen tube development can be achieved in artificial media devoid of a metabolizable source other than the compound under study. If *myo*-inositol is added, it functions as a precursor of uronic acid and pentose constituents of pollen tube polysaccharides (Kroh and Loewus, 1968). Details of our work on *myo*-inositol metabolism in germinating *L. longiflorum* pollen have been presented elsewhere (Loewus *et al.*, 1973). Only that aspect relating to the functional nature of the *myo*-inositol oxidation pathway will be considered here.

Schopfer *et al.* (1962) found that 2-O,C-methylene-*myo*-inositol (Fig. 11) could alter growth and development of *Schizosaccharomyces pombe*. The same compound also inhibited *myo*-inositol oxidation in the rat (Weinhold and Anderson, 1967). When the compound was added to medium used for germinating pollen, it blocked both germination and tube elongation. Fortunately, a concentration range was found in which germination was only slightly affected while tube elongation was reduced to about 30 percent of the control. Using this level of 2-O,C-methylene-*myo*-inositol (abbreviated MMO), there is a significant decrease in the incorporation of label from *myo*-inositol-2-³H into pectic substance of cytoplasm and wall. One such experiment is shown in Table 4. Pollen was incubated for 9 hours: first, 3 hours in pentaerythritol medium, then 3 hours in the same medium with MMO added, and finally 3 hours in fresh pentaerythritol containing labeled *myo*-inositol or D-glucose. In the presence of MMO, incorporation of label from *myo*-inositol-2-³H into tube wall polysaccharide was reduced to 26 percent of the control. Incorporation of label from D-glucose-1-¹⁴C revealed no apparent change.

Only when the distribution of label in hydrolyzed tube wall polysaccharide was examined was it possible to obtain a clear view of the MMO effect. In the case of *myo*-inositol-2-³H-labeled tubes, MMO repressed

TABLE 4

Effect of MMO on Incorporation of *myo*-Inositol-2-^{3}H and D-Glucose-1-^{14}C
into *Lilium longiflorum* Pollen

	Portion of radioisotope recovered (%)			
	myo-Inositol-2-^{3}H		D-Glucose-1-^{14}C	
Fraction	−MMO	+MMO	−MMO	+MMO
Cytoplasmic carbohydrate	8.5	4.4	32	32
Tube wall carbohydrate	10.5	2.7	34	34
Hydrolyzed fractions from tube wall			(Tube wall carbohydrate only)	
Uronic acid	++	+	7.5	2.5
Galactose + glucose	+++	−	76	92
Arabinose	++++	+	9	3
Xylose	+	−	1	0.5
Rhamnose	−	−	2	1

incorporation of label into products of *myo*-inositol oxidation, that is, uronic acid and pentose constituents. Further, since xylose biosynthesis was blocked, this pentose did not recycle to hexose and no label appeared in glucose or galactose residues (Loewus and Jang, 1958). In the case of D-glucose-1-^{14}C-labeled tubes, MMO did not block D-glucose incorporation into tube walls. Rather, it shifted the proportions of label incorporated through *myo*-inositol and UDP-D-glucose, diminishing the amount of ^{14}C in uronic acid and arabinose residues, and raising the amount shunted to hexose residues. The reduced level of tube wall rhamnose, which is not a product of *myo*-inositol oxidation, probably reflects a lower rate of galacturonorhamnan biosynthesis due to the block in galacturonic acid biosynthesis.

Our MMO studies provide evidence that the *myo*-inositol oxidation pathway has a dominant role in UDP-D-glucuronic acid biosynthesis in germinating *L. longiflorum* pollen. A similar conclusion has been reached in the case of cell wall polysaccharide biosynthesis in cultured sycamore cells (Rubery and Northcote, 1970). The effect of MMO in UDP-D-glucose dehydrogenase was not examined at the time these experiments were run, but a study currently in progress indicates that MMO does not inhibit this enzyme.

As evidence accumulates in favor of the functional existence of a *myo*-inositol oxidation pathway to UDP-D-glucuronic acid, the question of the quantitative role of this pathway emerges. At this time information is quite limited. Roberts (1971) found that UDP-D-glucuronic acid pyro-phosphorylase activity in barley seedlings kept pace with cell wall polysaccharide biosynthesis while UDP-D-glucose dehydrogenase activity was undetectable. Inability to detect the latter enzyme must be regarded with caution since Davies and Dickinson (1972) have found that release of this enzyme into solution under certain circumstances, for example, from ungerminated pollen, required addition of a nonionic detergent to the extracting medium. Rubery (1972) examined sycamore cambium and xylem tissue for UDP-D-glucose dehydrogenase and found the enzyme was significantly more active in xylem than in cambium. He assigned to this enzyme the biosynthetic role of producing precursor for xylan in secondary wall formation and to the *myo*-inositol oxidation pathway, a function related to primary wall biosynthesis. Roberts and Cetorelli (1973) surveyed a number of plants for both UDP-D-glucose dehydrogenase and UDP-D-glucuronic acid pyrophosphorylase activities. They concluded that those tissues where primary cell wall biosynthesis predominates probably synthesize UDP-D-glucuronic acid by the *myo*-inositol oxidation pathway. Further, using *Acer rubrum* as experimental tissue, they confirmed the work of Rubery which had implicated UDP-D-glucose dehydrogenase in the synthesis of UDP-D-glucuronic acid for polysaccharide biosynthesis in secondary cell wall processes. Even the work of Jung *et al.* (1972), in which a study of *myo*-inositol metabolism in a *myo*-inositol-requiring callus tissue culture of *Fraxinus pennsylvania* led to the conclusion that *myo*-inositol did not have a major role during pectin formation, may find its explanation in the fact that UDP-D-glucuronic acid formation in this particular tissue arises from UDP-D-glucose once primary wall substance has been laid down.

myo-Inositol as an Intermediate in Carbohydrate Metabolism and Sugar Interconversions

Insofar as is known, *myo*-inositol biosynthesis is limited to a single reaction in which D-glucose 6-phosphate is converted to *myo*-inositol 1-phosphate. In plants, experimental evidence is still unavailable as regards the stereospecific form of *myo*-inositol 1-phosphate produced, but in other tissues it is 1L-*myo*-inositol 1-phosphate. Beyond this synthetic step *myo*-inositol participates in a number of reactions, as outlined in the diagram in Fig. 12.

There appears to be a quantity of free *myo*-inositol in all plants. From this source, *myo*-inositol is withdrawn for resynthesis of the monophosphate and for each of the various secondary products associated with *myo*-inositol metabolism. It is unlikely that all metabolic events which require free

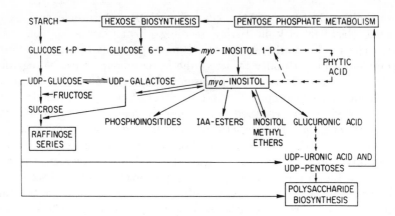

Fig. 12. A scheme of *myo*-inositol biosynthesis and metabolism in plants and the interrelationships between *myo*-inositol and other carbohydrate interconversions.

myo-inositol draw their *myo*-inositol requirement from a single pool. Rather, they probably complete for substrate through a highly regulated process. Some inkling of this has emerged in studies of phytic acid biosynthesis and metabolism where synthesis appears to be regulated by the level of each successively higher inositol polyphosphate and ADP, as well as by a protein inhibitor of the enzyme occurring in the same tissue (Majumder and Biswas, 1973a, b). Subsequent dephosphorylation, again a stepwise sequence, occurs in stereospecific fashion, highly regulated by the kind of phytase, the inorganic phosphate level, and the pattern of phytic acid localization. There is even a suggestion that one phytase may produce some *myo*-inositol 1-phosphate as well as the usual 2-phosphate (Lim and Tate, 1973; Tomlinson and Ballou, 1962).

The flow of hexose to uronic acid and related products of uronic acid metabolism shows every sign of being a highly regulated process in which alternate pathways, either through UDP-D-glucose or *myo*-inositol, determine the rate of synthesis of cell wall polysaccharides. Since UDP-D-glucose dehydrogenase is readily inhibited by UDP-D-xylose as well as UDP-D-galacturonic acid and UDP-D-glucuronic acid (Dickinson *et al.*, 1973), accumulation of these products in the plant cell from either pathway will probably repress UDP-D-glucose oxidation. Even the activity of UDP-D-glucose epimerase (Rubery, 1973) may be affected if a significant amount of free *myo*-inosose-2 is produced (Ketley and Schellenberg, 1972). The possible role of cycloaldolase in generating inosose has been discussed earlier in this chapter.

Methyl ethers of *myo*-inositol and other inositols are fairly common in plants. The suggestion has been made that certain of these ethers provide monomeric precursors of methylated polysaccharides (Loewus, 1964a). It is also possible, as indicated in the diagram, that these products are recalled into the *myo*-inositol pool by physiological events during plant growth and that in this process, methyl groups are released for biosynthetic requirements in other parts of the plant cell.

Although intensive study of indol-3-acetyl esters of *myo*-inositol has been restricted to corn, the occurrence of such compounds in plant tissue should prompt future research on its regulatory role. The peculiar association of indol-3-acetate, a plant hormone whose effects on cell wall structure are well established, with *myo*-inositol, another compound closely associated with the biosynthesis of cell wall polysaccharides, leads one to regard this group of esters as potentially important in the control of processes related to cell wall development.

Nothing has been said of phosphoinositides in this discussion but this should not be interpreted other than as a limitation placed on the content of this chapter. Klyashchitshii *et al.* (1969) have reviewed this subject from the viewpoint of structure, biochemistry, and synthesis. The obvious importance of these compounds in membrane interactions is well appreciated. Recently, Lim and Tate (1971) isolated a phospholipid activator of wheat bran phytase (F_1 fraction) and identified it as lysolecithin.

The role of *myo*-inositol as a cofactor in syntheses of the raffinose family of sugars provides a fine example of the intimate relationship between carbon flow through UDP-D-glucose to sucrose and inositol metabolism. At one end of the scheme, free *myo*-inositol is required for galactinol biosynthesis. The size of this free inositol pool will determine how much UDP-D-glactose is transferred to sucrose and higher homologs of the raffinose series. The rate of removal of UDP-D-galactose will probably influence carbon flow into sucrose. At the same time, other demands on both *myo*-inositol and UDP-D-glucose sources for polysaccharide biosynthesis at the other end of the scheme will alter these relationships. Although it is premature to speculate on the nature of carbohydrate regulation in this scheme, there is little doubt that a highly controlled process maintains this free inositol pool. One can no longer regard free *myo*-inositol as an inert secondary plant product. Research on *Lilium longiflorum* pollen (C-L. Rosenfield, unpublished studies) indicates that a significant portion of the carbon from *myo*-inositol finds its way back into hexose biosynthesis over a metabolic route that takes it through pentose. Thus, even the products of *myo*-inositol oxidation may ultimately be under a form of control determined by competition between polysaccharide biosynthesis and recycling processes.

ACKNOWLEDGMENT

This work was supported by grant GM-12422 from the National Institute of General Medical Sciences, National Institutes of Health.

REFERENCES

Anderson, L. 1966. *Annu. Rev. Plant Physiol.* **17**:209.
Anderson, L. 1972. *In* "The Carbohydrates" (W. Pigman and D. Horton, eds.), Vol. IA, pp. 519–579. Academic Press, New York.
Angyal, S. J., and A. F. Russell. 1969. *Aust. J. Chem.* **22**:383.
Bandurski, R. S., and Z. Piskornik. 1973. *In* "Biogenesis of Plant Cell Wall Polysaccharides" (F. Loewus, ed.), pp. 297–314. Academic Press, New York.
Barnett, J. E. G., and D. L. Corina. 1968. *Biochem. J.* **108**:125.
Barnett, J. E. G., R. E. Brice, and D. L. Corina. 1970. *Biochem. J.* **119**:183.
Barnett, J. E. G., A. Rasheed, and D. L. Corina. 1973. *Biochem. J.* **131**:21.
Blank, G. E., J. Pletcher, and M. Sax. 1971. *Biochem. Biophys. Res. Commun.* **44**:319.
Brown, D. M., and M. E. Tate. 1973. *Proc. Aust. Biochem. Soc.* **6**:44.
Brown, R. J., and R. F. Serro. 1953. *J. Amer. Chem. Soc.* **75**:1040.
Brunner, A., M. Z. Piña, V. Chagoya de Sánchez, and E. Piña. 1972. *Arch. Biochem. Biophys.* **150**:32.
Charalampous, F. C. 1957. *J. Biol. Chem.* **225**:595.
Charalampous, F. C. 1960. *J. Biol. Chem.* **235**:1286.
Chen, I-W., and F. C. Charalampous. 1964. *Biochem. Biophys. Res. Commun.* **17**:521.
Chen, I-W., and F. C. Charalampous. 1966. *J. Biol. Chem.* **241**:194.
Chen, I-W., and F. C. Charalampous. 1967. *Biochim. Biophys. Acta* **136**:568.
Chen, M-S. 1972. Ph.D. Thesis, State University of New York, Buffalo.
Cosgrove, D. J. 1966a. *Rev. Pure Appl. Chem.* **16**:209.
Cosgrove, D. J. 1966b. *J. Sci. Food Agr.* **17**:550.
Cosgrove, D. J. 1970. *Aust. J. Biol. Sci.* **23**:1207.
Daughaday, W. H., J. Larner, and C. Hartnett. 1955. *J. Biol. Chem.* **212**:869.
Davies, M. D. and D. B. Dickinson. 1972. *Arch. Biochem. Biophys.* **152**:53.
Dickinson, D. B., J. E. Hopper, and M. D. Davies. 1973. *In* "Biogenesis of Plant Cell Wall Polysaccharides" (F. Loewus, ed.), pp. 29–48. Academic Press, New York.
Dietz, M., and P. Albersheim. 1965. *Biochem. Biophys. Res. Commun.* **19**:598.
Dittrich, P. and O. Kandler. 1972. *Phytochemistry* **11**:1729.
Dittrich, P., M. Gietl, and O. Kandler. 1972. *Phytochemistry* **11**:245.
Eagle, H., B. W. Agranoff, and E. E. Snell. 1960. *J. Biol. Chem.* **235**:1891.
Ehmann, A., and R. S. Bandurski. 1972. *J. Chromatogr.* **72**:61.
Ehmann, A., and R. S. Bandurski. 1973. *Plant Physiol.* **51**:S-12.
Eisenberg, F., Jr. 1966. *In* "Cyclitols and Phosphoinositides" (H. Kindl, ed.), Vol. 2, Proc. 2nd Meet., FEBS. pp. 3–13. Pergamon, Oxford.
Eisenberg, F., Jr. 1967. *J. Biol. Chem.* **242**:1375.
Eisenberg, F., Jr. 1969. *Ann. N. Y. Acad. Sci.* **165**:509.
Eisenberg, F., Jr., and A. H. Bolden. 1965. *Biochem. Biophys. Res. Commun.* **21**:100.
Eisenberg, F., Jr. and A. H. Bolden. 1968. *Fed. Proc., Fed. Amer. Soc. Exp. Biol.* **27**:595.
Eisenberg, F., Jr., A. H. Bolden, and F. Loewus. 1964. *Biochem. Biophys. Res. Commun.* **14**:419.

Feingold, D. S., and D-F. Fan. 1973. *In* "Biogenesis of Plant Cell Wall Polysaccharides" (F. Loewus, ed.), pp. 69–84. Academic Press, New York.

Frydman, R. B., and E. F. Neufeld. 1963. *Biochem. Biophys. Res. Commun.* **12**:121.

Halliday, J. W., and L. Anderson. 1955. *J. Biol. Chem.* **217**:797.

Hauska, G., and O. Hoffmann-Ostenhof. 1967. *Hoppe-Seyler's Z. Physiol. Chem.* **348**: 1558.

Hoffmann, H., I. Wagner, and O. Hoffmann-Ostenhof. 1969. *Hoppe-Seyler's Z. Physiol. Chem.* **350**:1465.

Hoffmann-Ostenhof, O., C. Jungwith, and I. B. Dawid. 1958. *Naturwissenschaften* **45**:265.

Horecker, B. L., O. Tsolas, and C. Y. Lai. 1972. *In* "The Enzymes" (P. D. Boyer, ed.), 3rd ed., Vol. 7, pp. 213–258. Academic Press, New York.

IUPAC Commission on the Nomenclature of Organic Chemistry and IUPAC-IUB Commission on Biochemical Nomenclature. 1968. *J. Biol. Chem.* **243**:58–59.

Imhoff, V. 1970. *C. R. Acad. Sci.* **270**:2441.

Imhoff, V., and R. Bourdu. 1970. *Physiol. Veg.* **8**:649.

Imhoff, V., and R. Bourdu. 1973. *Phytochemistry* **12**:331.

Irving, G. C. J., and D. J. Cosgrove. 1972. *J. Bacteriol.* **112**:434.

Johnson, L. F., and M. E. Tate. 1969a. *Can. J. Chem.* **47**:63.

Johnson, L. F., and M. E. Tate. 1969b. *Ann. N. Y. Acad. Sci.* **165**:526.

Jones, R. 1969. *Planta* **85**:359; **87**:119; **88**:73.

Jung, P., W. Tanner, and K. Wolter. 1972. *Phytochemistry* **11**:1655.

Ketley, J. N., and K. A. Schellenberg. 1972. *Biochim. Biophys. Acta* **284**:549.

Kiely, D. E., and H. G. Fletcher, Jr. 1969. *J. Org. Chem.* **34**:1386.

Kindl, H., ed. 1966a. "Cyclitols and Phosphoinositides," Vol. 2. Proc. 2nd Meet. FEBS. Pergamon, Oxford.

Kindl, H. 1966b. *In* "Cyclitols and Phosphoinositides" (H. Kindl, ed.), Vol. 2, Proc. 2nd Meet. FEBS. pp. 15–22. Pergamon, Oxford.

Kindl, H., and O. Hoffmann-Ostenhof. 1964a. *Biochem. Z.* **339**:374.

Kindl, H., and O. Hoffmann-Ostenhof. 1966a. *Progr. Chem. Org. Natur. Prod.* **24**:149.

Kindl, H., and O. Hoffmann-Ostenhof. 1966b. *Hoppe-Seyler's Z. Physiol. Chem.* **345**:257.

Kindl, H., and O. Hoffmann-Ostenhof. 1966c. *Monatsh. Chem.* **97**:1778.

Kindl, H., and O. Hoffmann-Ostenhof. 1966d. *Monatsh. Chem.* **97**:1771.

Klyashchitskii, B. A., S. D. Sokolov, and V. I. Shvets. 1969. *Russ. Chem. Rev.* **38**:345.

Kremlicka, G. J., and O. Hoffmann-Ostenhof. 1966. *Hoppe-Seyler's Z. Physiol. Chem.* **344**:261.

Kroh, M., and F. Loewus. 1968. *Science* **160**:1352.

Kurasawa, H., T. Hayakawa, and S. Motoda. 1967. *Agr. Biol. Chem.* **31**:382.

Labarca, C., P. B. Nicholls, and R. S. Bandurski. 1965. *Biochem. Biophys. Res. Commun.* **20**:641.

Lai, C. Y., and B. L. Horecker. 1972. *Essays Biochem.* **8**:149–178.

Lim, P. E., and M. E. Tate. 1971. *Biochim. Biophys. Acta* **250**:155.

Lim, P. E., and M. E. Tate. 1973. *Biochim. Biophys. Acta* **302**:316.

Loewus, F. 1964a. *Nature (London)* **203**:1175.

Loewus, F. 1964b. *Arch. Biochem. Biophys.* **105**:590.

Loewus, F. 1965. *Fed. Proc., Fed. Amer. Soc. Exp. Biol.* **24**:855.

Loewus, F. 1969. *Ann. N. Y. Acad. Sci.* **165**:577.

Loewus, F. 1971. *Annu. Rev. Plant Physiol.* **22**:337.

Loewus, F., and R. Jang. 1958. *J. Biol. Chem.* **232**:521.

Loewus, F., and S. Kelly. 1962. *Biochem. Biophys. Res. Commun.* **7**:204.

Loewus, F., S. Kelly, and E. F. Neufeld. 1962. *Proc. Nat. Acad. Sci. U.S.* **48**:412.

Loewus, F., M-S. Chen, and M. W. Loewus. 1973. *In* "Biogenesis of Plant Cell Wall Polysaccharides" (F. Loewus, ed.), pp. 1–27. Academic Press, New York.

Loewus, M. W., and F. Loewus. 1971. *Plant Physiol.* **48**:255.

Loewus, M. W., and F. Loewus. 1973a. *Plant Physiol.* **51**:263.

Loewus, M. W., and F. Loewus. 1973b. *Plant Sci. Lett.* **1**:65.

Majumder, A. L., and B. B. Biswas. 1973a. *Phytochemistry* **12**:315.

Majumder, A. L., and B. B. Biswas. 1973b. *Phytochemistry* **12**:321.

Majumder, A. L., N. C. Mandel, and B. B. Biswas. 1972. *Phytochemistry* **11**:503.

Malangeau, P. 1958. *Qual. Plant. Mater. Veg.* **3–4**:393.

Mandel, N. C., and B. B. Biswas. 1970. *Indian J. Biochem.* **7**:63.

Maquenne, L. 1900. "Les sucres et leurs principaux derives," p. 189. Paris.

Matheson, N. K., and M. St. Clair. 1971. *Phytochemistry* **10**:1299.

Matheson, N. K., and S. Strother. 1969. *Phytochemistry* **8**:1349.

Mogyoros, M., A. Brunner, and E. Piña. 1972. *Biochim. Biophys. Acta* **289**:420.

Moore, T. S. 1973. *Plant Physiol.* **51**:529.

Moreno, A., and C. Cardini. 1966. *Plant Physiol.* **41**:909.

Nicholls, P. B. 1967. *Planta* **72**:258.

Nicholls, P. B., B. L. Ong, and M. E. Tate. 1971. *Phytochemistry* **10**:2207.

Piña, E., and E. L. Tatum. 1967. *Biochim. Biophys. Acta* **136**:265.

Piña, E., Y. Saldaña, A. Brunner, and V. Chagoya. 1969. *Ann. N. Y. Acad. Sci.* **165**:541.

Piskornik, Z., and R. S. Bandurski. 1972. *Plant Physiol.* **50**:176.

Plouvier, V. 1963. *In* "Chemical Plant Taxonomy" (T. Swain, ed.), pp. 313–336. Academic Press, New York.

Posternak, T. 1965. "The Cyclitols." Holden-Day, San Francisco, California.

Roberts, R. M. 1971. *J. Biol. Chem.* **246**:4995.

Roberts, R. M., and J. J. Cetorelli. 1973. *In* "Biogenesis of Plant Cell Wall Polysaccharides" (F. Loewus, ed.), p. 49–68. Academic Press, New York.

Roberts, R. M., and F. Loewus. 1973. *Plant Physiol.* **52**:646.

Rubery, P. H. 1972. *Planta* **103**:188.

Rubery, P. H. 1973. *Planta* **111**:267.

Rubery, P. H., and D. H. Northcote. 1970. *Biochim. Biophys. Acta* **222**:95.

Ruis, H., E. Molinari, and O. Hoffmann-Ostenhof. 1967. *Hoppe-Seyler's Z. Physiol. Chem.* **348**:1705.

Schilling, N., P. Dittrich, and O. Kandler. 1972. *Phytochemistry* **11**:1401.

Scholda, R., G. Billek, and O. Hoffmann-Ostenhof. 1964a. *Monatsh. Chem.* **95**:1305.

Scholda, R., G. Billek, and O. Hoffmann-Ostenhof. 1964b. *Monatsh. Chem.* **95**:1311.

Scholda, R., G. Billek, and O. Hoffmann-Ostenhof. 1964c. *Hoppe-Seyler's J. Physiol. Chem.* **335**:180.

Scholda, R., G. Billek, and O. Hoffmann-Ostenhof. 1964d. *Hoppe-Seyler's Z. Physiol. Chem.* **337**:277.

Schopfer, W. H., T. Posternak, and D. Wustenfeld. 1962. *Arch. Mikrobiol.* **44**:113.

Sensor, M., and O. Kandler. 1967. *Phytochemistry* **6**:1533.

Sherman, W. R., M. A. Stewart, P. C. Simpson, and S. L. Goodwin. 1968. *Biochemistry* **7**:819.

Sherman, W. R., M. A. Stewart, and M. Zinbo. 1969. *J. Biol. Chem.* **244**:5703.

Sherman, W. R., S. L. Goodwin, and K. D. Gunnell. 1971. *Biochemistry* **10**:3491.

Stinson, E. E., C. J. Dooley, J. M. Purcell, and J. S. Ard. 1967. *J. Agr. Food Chem.* **15:** 394.

Tanner, W. 1967. *Ber. Deut. Bot. Ges.* **80:**592.

Tanner, W., and O. Kandler. 1968. *Eur. J. Biochem.* **4:**233.

Tomlinson, R. V., and C. E. Ballou. 1962. *Biochemistry* **1:**166.

Ueda, M., A. Ehmann, and R. S. Bandurski. 1970. *Plant Physiol.* **45:**715.

Wagner, I., H. Hoffmann, and O. Hoffmann-Ostenhof. 1969. *Hoppe-Seyler's Z. Physiol. Chem.* **350:**1460.

Webb, J. A. 1973. *Plant Physiol.* **52:**S-12.

Weinhold, P. A., and L. Anderson. 1967. *Arch. Biochem. Biophys.* **122:**529.

UNUSUAL FATTY ACIDS IN PLANTS

T. GALLIARD

Agricultural Research Council, Food Research Institute,
Norwich, England

Introduction.. 209
Structure and Occurrence of Unusual Fatty Acids........................... 210
  Major (Usual) Fatty Acids.. 210
  Fatty Acid Composition of Plant Lipids................................. 211
  Occurrence of Unusual Fatty Acids in Higher Plants..................... 222
Unusual Fatty Acids and Chemotaxonomy.................................... 227
Biosynthesis of Oxygenated Fatty Acids..................................... 228
  2-Hydroxy Acids.. 229
  ω-Hydroxy and ω-Dicarboxylic Fatty Acids............................... 229
  In-Chain Monohydroxy Nonconjugated Fatty Acids........................ 230
  Epoxy and vic-Dihydroxy Fatty Acids.................................... 231
  Lipoxygenase-Mediated Reactions and Fatty Acids with Conjugated Unsaturation.. 232
  Chain Length Modifications... 236
Regulatory Factors and Unusual Fatty Acids................................. 237
  Environmental Aspects.. 237
  Structural and Genetic Factors.. 238
  Developmental Aspects.. 238
  Control at the Cellular Level.. 238
References.. 239

Introduction

This review does not attempt comprehensive coverage of the subject area. The discussion is limited to the flowering plants, and treatment of lower plants, yeasts, and photosynthetic bacteria is omitted. Furthermore, emphasis will be given to oxygenated fatty acids and discussion of unusual nonoxygenated fatty acids will be restriced to cases where these are of direct relevance to corresponding oxygenated derivatives. Recent reviews have comprehensively covered the general field of the occurrence (Wolff, 1966; Smith, 1970; Hitchcock and Nichols, 1971; Pohl and Wagner, 1972a,b) and biochemistry (Hitchcock and Nichols, 1971) of unusual fatty acids in plants.

A liberal interpretation of "unusual" is used and the discussion will cover fatty acids that are considered unusual because (a) they occur as major components of lipids in only certain plants, (b) although they may occur widely, knowledge of their structures and biochemistry has only recently been demonstrated, or (c) they do not occur in significant amounts in healthy tissues but are formed on cell disruption. Fatty acids which may occur widely in plants but only in trace amounts may be considered as minor fatty acids and are not discussed here.

Structure and Occurrence of Unusual Fatty Acids

MAJOR (USUAL) FATTY ACIDS

The normal and ubiquitous major fatty acids of plants are relatively few in number and at least 90 percent of the total fatty acids of most plant tissues are represented by one saturated acid, palmitic (16:0)* and three unsaturated C_{18} acids—oleic (18:1, ω9), linoleic (18:2, ω6,9), and linolenic (18:3, ω3,6,9) acids. The saturated acids—lauric (12:0), myristic (14:0), and stearic (18:0)—rarely occur at more than a few percent of the total fatty acids. The C_{16} homolog of linolenic acid (16:3, ω3,6,9) is an important component of the leaf lipids of many angiosperms of several different orders (Jamieson and Reid, 1971b).

---

* The abbreviated notation for fatty acids follows a conventional system (see Hitchcock and Nichols, 1971) in which the figures preceding the colon represent the number of carbon atoms, the figure immediately following the colon refers to the number of unsaturated bonds, and the position and configuration of unsaturated centers are indicated by further symbols. The biochemical relationships between different fatty acids are, in many cases, clarified when functional groups are numbered from the methyl terminal group and such numbers are always prefixed with ω; omission of a symbol indicates conventional numbering from the carboxyl end but where confusion may arise the $\triangle$ symbol is used in these cases. The letters c and t refer to cis and trans ethylenic bonds, respectively; e refers to a terminal ethylenic group or one of unknown configuration; a represents an acetylenic group. Omission of a symbol for ethylenic configuration infers a cis bond.

FATTY ACID COMPOSITION OF PLANT LIPIDS

Fatty acids rarely occur at significant levels in the free acid form in healthy plant tissues, but are present in complexed forms as lipids. In most cases, fatty acids are bound through a carboxyl ester as acyl lipids, although amide and ether linkages are formed in a few lipids. In an overall generalization, the majority of acyl lipids of plant tissues may be divided into three types: somatic, storage, and surface lipids.

Somatic Lipids

The phospholipids and glycolipids of lipoprotein membrane structures represent the major location of acyl lipids in most types of plant cells. The reported occurrence of unusual fatty acids in such lipids is uncommon. In oil seeds with unusual fatty acids in the storage lipid triglycerides, the membrane lipids generally contain only the normal fatty acids (Vijayalakshmi and Rao, 1972). However, exceptions to this generalization have recently been noted; cyclopropane fatty acids (Kaimal and Lakshminarayana, 1972) or both cyclopropane and cyclopropene acids (Yano *et al.*, 1972) occur in somatic polar lipids as well as in the seed oils of plants in the family Malvaceae.

Kuiper and Stuiver (1972) have recently shown high levels of cyclopropane acids in the sulfolipid of early flowering plants and in the phosphatidylcholine of drought-tolerant plants. The unusual ethylenic acids, γ-linolenic acid and stearidonic acid, are found in the galactolipids of leaves from some plants in the families Boraginaceae and Caryophyllaceae (Jamieson and Reid, 1969, 1971a).

In contrast to seed triglyceride analyses, careful studies on the isolation and characterization of the polar lipids in plants have been performed only recently and it must be assumed that the list of unusual fatty acids in somatic lipids will be extended.

Storage Lipids

Triglycerides represent the major lipid storage form in plants and a high proportion of the known unusual fatty acids occur as acyl components of triglycerides, particularly in seed oils (see Tables 1–3). Less is known about the fatty acid compositions of the triglycerides from fruit exocarp or of the lipid droplets in cells of other tissues; from the limited amount of information available to date, there is nothing unusual about the fatty acids of these lipids (Hitchcock and Nichols, 1971).

Surface Lipids

The cuticular surface of plants comprises a network of cross-esterified oxygenated fatty acids (the cutin) embedded in a hydrophobic layer,

TABLE 1

Oxygenated Forms of Saturated Fatty Acids Occurring Naturally in Higher Plants

Fatty acid carbon chain length	Type and position of oxygenated group	Trivial name of fatty acid	Typical sources	References[a]
16–26	2-Hydroxy	—	Widely occurring in cerebrosides and phytoglycolipid	
14	9-Hydroxy	—	—	
14	10-Hydroxy	—	Leaf cutin from *Coffea arabica*	Holloway *et al.* (1972)
18	10-Hydroxy	—	Orujo (sulphur olive oil)	*
22	13-Hydroxy	Phellonic	Tall oil	*
16	16-Hydroxy	—	Cutins and suberins	Kolattukudy and Walton (1972), Holloway and Deas (1973)
18	18-Hydroxy	—	Cutins	*
18	*cis*-9,10-Epoxy	—	Seeds of *Onguekoa gore* (Olacaceae) and *Tragopogon porrifolius* (Compositae)	*
18	*threo*-9,10-Dihydroxy	—	Seeds of *Onguekoa gore* (Olacaceae)	*
18	*erythro*-9,10-(ω9,10)-Dihydroxy	—	Seeds of *Ricinus communis* (Euphorbiaceae)	*
20	*erythro*-11,12(ω9,10)-Dihydroxy	—	Seeds of *Cardamine impatiens* (Cruciferae)	*
22	*erythro*-13,14(ω9,10)-Dihydroxy	—	Seeds of *Cardamine impatiens* (Cruciferae)	*
24	*erythro*-15,16(ω9,10)-Dihydroxy	—	Seeds of *Cardamine impatiens* (Cruciferae)	*

16	9,16-Dihydroxy	—	} Cutins	Kolattukudy and Walton (1972), Holloway et al. (1972)
16	10,16-Dihydroxy	—		
16	9-Hydroxy-16-oxo	—	Cutin from embryonic *Vicia faba*	Kolattukudy (1972)
18	10,18-Dihydroxy	—	Cutin from *Agave americana* leaf	*
18	9,10-Epoxy-18-hydroxy	—	Cutins and suberins	Kolattukudy and Walton (1972), Holloway and Deas (1973)
18	9,10-Epoxy-18-oxo	—	Cutin from immature fruit of apple *Pirus malus* seed of *Champaepeuce afra* (Compositae)	Kolattukudy (1973)
18	9,10,18-Trihydroxy	Phloionolic	Cutins and suberins	Kolattukudy and Walton (1972), Holloway and Deas (1973)
10	7-Hydroxy	—	} Present as glycosides in *Ipomoea* spp. (Convolulaceae)	*
14	11-Hydroxy	Convolvulinolic		
16	11-Hydroxy	Jalapinolic		
14	3,11-Dihydroxy	—		
16	3,12-Dihydroxy	—		

a This table is compiled in part from data given by Smith (1970) and Hitchcock & Nichols (1971), and further information and full literature citations on compounds marked * may be obtained from these reviews.

TABLE 2

Oxygenated and Related Nonoxygenated Fatty Acids with Nonconjugated Ethylenic Bonds, Naturally Occurring in Higher Plants

Chain length and No. of ethylenic centers	Position(s) of oxygenated function(s)	Position(s) of ethylenic center(s) (cis unless t indicated)	Trivial name of fatty acid	Typical sources	References[a]
18:1	—	ω9(Δ9)	*Oleic*	Ubiquitous	—
20:1	—	ω9(Δ11)	Asclepic	Seed of *Cardiospermum halicacabum* (Sapindaceae)	*
22:1	—	ω9(Δ13)	Erucic	Seeds of various genera (Cruciferae)	*
24:1	—	ω9(Δ15)	Nervonic	Seeds of *Lunaria biennis* (Cruciferae) and *Ximenia americana* (Sapindaceae)	*
26:1	—	ω9(Δ17)	Ximenic	} Seeds of *Ximenia* spp. (Sapindaceae)	*
28:1	—	ω9(Δ19)	—	
30:1	—	ω9(Δ21)	Lumequeic	
16:1	Δ12-Hydroxy	ω9(Δ7)	—	Seed of *Lesquerella densipata* (Cruciferae)	*
18:1	Δ12(ω7)-Hydroxy	ω9(Δ9)	Ricinoleic	Seeds of *Ricinus* spp. (Euphorbiaceae)	*
20:1	ω7-Hydroxy	ω9(Δ11)	Lesquerolic	Seed of *Lesquerella lasiocarpa* (Cruciferae)	*

18:1	(+)-Δ12,13(ω6,7)-Epoxy	ω9(Δ9)	(+)-Vernolic	Seeds of *Vernonia anthelmintica* (Compositae) and some other families	Kolattukudy and Walton (1972)	*
18:1	(−)-Δ12,13(ω6,7)-Epoxy	ω9(Δ9)	(−)-Vernolic	Seed of *Malope trifida* (Malvaceae)		*
18:1	Δ18(ω1)-Hydroxy	ω9(Δ9)	—	Cutins		*
18:1	Δ18(ω1)-Hydroxy	ω9(Δ9)t	—	Suberin (cork) of *Quercus suber*		—
18:2	—	ω6,9(Δ9,12)	*Linoleic*	Ubiquitous		*
20:2	—	ω6,9(Δ11,14)	—	Seed of *Ephedra camylopoda* (Ephedraceae)		*
18:2	Δ18(ω1)-Hydroxy	ω6,9(Δ9,12)	—	Cutins	Kolattukudy and Walton (1972)	—
18:2	Δ15,16(ω3,4)-Epoxy	ω6,9(Δ9,12)	—	Seed of *Camelina sativa* (Cruciferae)		*
18:2	—	ω6(Δ12)t,ω9(Δ9)	—	Seeds of *Dimorphoteca sinuata* and *Crepis rubra* (Compositae)		*
18:2	—	ω6(Δ12)t,ω9(Δ9)t	—	Seed of *Chilopsis linearis* (Bignoniaceae)		*
18:3	—	ω3,6,9(Δ9,12,15)	*α-Linolenic*	Ubiquitous		—
16:3	—	ω3,6,9(Δ7,10,13)	—	Galactolipids of leaves from many, but not all, plants	Jamieson and Reid (1971b)	—
17:3	—	ω3,6,9(Δ8,11,14)	Norlinolenic	Seed of *Thymus vulgaris* (Labitae)		*
18:3	Δ2-Hydroxy	ω3,6,9(Δ9,12,15)	α-Hydroxylinolenic	Seed of *Thymus vulgaris* (Labitae)		*

(Continued)

TABLE 2—(Continued)

Chain length and No. of ethylenic centers	Position(s) of oxygenated function(s)	Position(s) of ethylenic center(s) (cis unless t indicated)	Trivial name of fatty acid	Typical sources	References[a]
20:3	—	ω3,6,9(Δ11,14,17)	—	Seed of *Ephedra campylopoda* (Ephedraceae)	*
18:2	ω7(Δ12)-Hydroxy	ω3,9(Δ9,15)	Densipolic	Seed of *Lesquerella densipata* (Cruciferae)	*
20:2	ω7(Δ14)-Hydroxy	ω3,9(Δ9,15)	Auricolic	Seed of *Lesquerella ariculata* (Cruciferae)	Kleiman *et al.* (1972)
18:1	Δ9,10(ω9,10)-Epoxy	ω6(Δ12)	Coronaric	Seed of *Chrysanthemum coronarium* (Compositae)	*
18:1	Δ9,10-Epoxy-Δ18(ω1)-hydroxy	ω6(Δ12)	—	Cutins	Kolattukudy and Walton (1972), Holloway and Deas (1973)

18:1	Δ9,10,18-Trihydroxy	$\omega6(\Delta12)$	—	Seeds of *Champaepeuce afra* (Compositae) and cutins	Kolattukudy and Walton (1972)	*
24:1	$\omega10(\Delta15)$-Oxo	$\omega6(\Delta18)$	—			
26:1	$\omega10(\Delta17)$-Oxo	$\omega6(\Delta20)$	—	Seeds of *Cuspidaria pterocarpa* (Bignoniaceae)		*
28:1	$\omega10(\Delta19)$-Oxo	$\omega6(\Delta22)$	—			
18:1	—	$\omega12(\Delta6)$	Petroselenic	Seeds of various genera (Umbelliferae and Araliaceae)		*
18:3	—	$\omega6,9,12(\Delta6,9,12)$	γ-Linolenic	Seeds and leaf galactolipids of various general (Boraginaceae, Caryophyllaceae, Onagraceae)	Jamieson and Reid (1969, 1971a)	
18:4	—	$\omega3,6,9,12(\Delta6,9,12,15)$	Stearidonic			
18:2	Δ9,10(ω9,10)-Epoxy	$\omega15(\Delta3)l,\omega6(\Delta12)$	—	Seeds of *Stenachaenium* spp. (Compositae)	Kleiman *et al.* (1971)	*
18:3	—	$\omega15(\Delta3)l,\omega6,9(\Delta9,12)$	—			

^{a} See footnote to Table 1.

TABLE 3

OXYGENATED AND RELATED NONOXYGENATED FATTY ACIDS WITH ACETYLENIC AND/OR CONJUGATED ETHYLENIC BONDS, NATURALLY OCCURRING IN HIGHER PLANTS

Fatty acid carbon chain length	Position of oxygenated function	Position of and configuration of unsaturated centers	Trivial name	Typical source (seed oil unless stated otherwise)	References[a]
6-Unsaturated Group					
10	—	ω6c,8t(Δ2t,4c)	—	*Sapium sebeferum* (Euphorbiaceae)	*
12	—	ω8c,10e(Δ2e,4e)	—	*Sebastiana lingustrina* (Euphorbiaceae)	*
18	ω10(Δ9)-Hydroxy	ω6c,8t(Δ10t,12c)	—	*Calendula officinalis* and *Xeranthemum annuum* (Compositae)	*
18	—	ω6t,8t(Δ10t,12t)	Jacaric	*Jarcaranda mimosifolia* (Bignoniaceae)	*
18	ω10(Δ9)-Hydroxy	ω6t,8t(Δ10t,12t)	Dimorphecolic	*Dimorpholeca sinuata* (Compositae)	*
18	ω10(Δ9)-Oxo	ω6t,8t(Δ10t,12t)	—	*Dimorpholheca sinuata* (Compositae)	*
18	—	ω6t,8t,10c(Δ8c,10t,12t)	Calendic	*Calendula officinalis* (Compositae)	*
18	—	ω6a,(Δ12a)	Tariric	*Pictannia* spp. (Olacaceae)	*
18	—	ω6a,9c(Δ9c,12a)	Crepenynic	*Crepis* spp. (Compositae) *Acanthosymis spinescens* (Santalaceae)	*

| 18 | — | ω4c, 6a, 9c(Δ9c, 12a, 14c) | Dehydrocrepenynic | *Afzelia cuanzensis* (Leguminosae) | * |
| 18 | ω10(Δ9)-Hydroxy | ω6a, 8(Δ10t, 12a) | Helenynolic | *Helichrysum bracteatum* (Compositae) | * |

9-Unsaturated Group

18	—	ω5c, 7t, 9c(Δ9c, 11t, 13c)	Punicic (trichosanic)	*Punicum granetum* (Punicaceae) *Trichosanthes cucumoides* (Cucurbitaceae)	*
18	—	ω5t, 7t, 9c(Δ9c, 11t, 13t)	α-Eleostearic	*Aleurites fordii* (Euphorbiaceae)	*
18	—	ω3c, 5t, 7t, 9c(Δ9c, 11t, 13t, 15c)	α-Pinaric	*Impatiens edgeworthii* (Balsaminaceae) *Parinarium laurinum* (Rosaceae)	*
18	ω6(Δ13)DR-Hydroxy	ω7t, 9c(Δ9c, 11t)	DR-Coriolic, artemesic	*Coriaria nepalensis* (Coriariaceae) *Xeranthemum annuum* (Compositae)	*
18	ω6(Δ13)LS-Hydroxy	ω7t, 9c(Δ9c, 11t)	LS-Coriolic	*Monnia emarginata* (Polygalaceae)	Phillips and Smith (1972)
18	ω1(Δ18)-Hydroxy	ω5t, 7t, 9c(Δ9c, 11t, 13t)	α-Kamlolenic	*Mallotus phillipinensis* (Euphorbiaceae)	*
18	ω15(Δ3)-Oxo	ω5t, 7t, 9c(Δ9c, 11t, 13t)	α-Licanic, coupeic	*Licania rigida* (Rosaceae)	*
18	ω15(Δ3)-Oxo	ω3c, 5t, 7t, 9c(Δ9c, 11t, 13t, 15c)	—	*Chrysobalanus icaco* (Rosaceae)	*
18	ω6(Δ13)-Oxo	ω7t, 9t(Δ9t, 11t)	—	*Monnina emarginata* (Polygalaceae)	Phillips and Smith (1972)
18	—	ω5c, 7t, 9t(Δ9t, 11t, 13c)	Catalpic	*Catalpa ovata* (Bignoniaceae)	*
18	—	ω9a(Δ9a)	Stearolic	various genera (Santalaceae)	*

(Continued)

TABLE 3—(Continued)

Fatty acid carbon chain length	Position of oxygenated function	Position of and configuration of unsaturated centers	Trivial name	Typical source (seed oil unless stated otherwise)	References[a]
18	—	ω1e,9a(Δ9a)	—	*Acanthosyris spinescens* (Santalaceae)	*
18	—	ω1e,7a,9a(Δ9a,11a,17e)	Isanic	*Onguekoa gore* (Olacaceae)	*
18	ω11(Δ8)-Hydroxy	ω1e,7a,9a(Δ9a,11a,17e)	Isanolic	*Onguekoa gore* (Olacaceae)	*
17	—	ω7t,9a(Δ8a,10t)	Pyrulic	*Pyrularia pubera* and *Acanthosyris spinescens* (Santalaceae)	*
18	—	ω7t,9a(Δ9a.11t)	Ximenynic	Various genera (Santalaceae) and *Ximenia* spp. (Olacacae)	*
17	ω11(Δ7)-Hydroxy	ω7t,9a(Δ8a,10t)	—	*Acanthosyris spinescens* (Santalaceae)	*
18	ω11(Δ8)-Hydroxy	ω7t,9a(Δ9a,11t)	Ximenynolic	*Ximenia caffia* (Olacaceae)	*
17	—	ω1e,7t,9a(Δ8a,10t,16e)	—	*Acanthosyris spinescens* (Santalaceae)	*

17	ω11(Δ7)-Hydroxy	ω1e,7l,9a(Δ8a,10l,16e)	—	*Acanthosyris spinescens* (Santalaceae)	*
18	—	ω1e,7l,9a(Δ9a,11t,17e)	—	*Pyrularia pubuera* (Santalaceae)	*
18	ω11(Δ8)-Hydroxy	ω1e,7l,9a(Δ9a,11t,17e)	—	*Pyrularia pubera* and *Acanthosyris spinescens* (Santalaceae)	*
18	—	ω5t,7l,9a(Δ9a,11t,13t)	—	*Ximenia americana* (Olacaceae)—wood	*
18	—	ω7a,9a(Δ9a,11a)	—	*Onguekoa gore* (Olacaceae)	*
18	ω11(Δ8)-Hydroxy	ω7a,9a(Δ9a,11a)	—	*Onguekoa gore* (Olacaceae)	*
18	—	ω5c,7a,9a(Δ9a,11a,13c)	Bolekic	*Onguekoa gore* (Olacaceae)	*
18	ω11(Δ8)-Hydroxy	ω5c,7a,9a(Δ9a,11a,13c)	—	*Onguekoa gore* (Olacaceae)	*
18	—	ω1e,5c,7a,9a(Δ9a,11a,13c,17e)	—	*Onguekoa gore* (Olacaceae)	*
18	ω11(Δ8)-Hydroxy	ω1e,5c,7a,9a(Δ9a,11a,13c,17e)	—	*Onguekoa gore* (Olacaceae) *Buckleya distichophylla* (Santalaceae)	*
18	—	ω5t,7a,9a(Δ9a,11a,13t)	Exocarpic	*Exocarpus cupressiformis* (Santalaceae)—root	*

[a] See footnote to Table 1.

commonly known as the wax. Strictly, a wax is an ester of a long-chain fatty acid with a long-chain alcohol; the waxy layer of cuticles is in fact a complex mixture of wax esters, hydrocarbons, fatty alcohols, ketones, and acids together with terpenoids and other components. Although several components of this waxy layer are related to, and derived from fatty acids, they will not be further treated here. Excellent reviews on the chemistry and biochemistry of cuticular lipids have been written recently by Mazliak (1968) and Kolattukudy and Walton (1972). The fatty acids of cutins and suberins, which are closely related with respect to fatty fatty acid composition (Holloway and Deas, 1973), will be discussed in the present review because they are similar to some of the unusual fatty acids in other tissues and because recent developments in characterization techniques and biochemical studies have increased interest in this area. The fact that the oxygenated fatty acids of cutins and suberins occur widely in the plant kingdom necessitates the liberal definition of "unusual" fatty acids.

OCCURRENCE OF UNUSUAL FATTY ACIDS IN HIGHER PLANTS

Explanation of Tables 1, 2, and 3

Tables 1, 2, and 3 were compiled mainly from information in recent definitive reviews by Smith (1970) and Hitchcock and Nichols(1971). Literature citations are only given for recent publications not included in the above reviews and the reader is referred to the above works for relevant references and further information. The order of listing in the tables is unconventional from a chemical point of view because an attempt is made to emphasize biochemical and phylogenetic relationships where appropriate. Thus fatty acids are not necessarily listed in order of increasing chain length, nor are substituted fatty acids treated separately from the paraffinic types. The separation into derivatives of saturated, nonconjugated ethylenic and acetylenic plus conjugated fatty acids is, even so, somewhat arbitrary since, for example, 9, 10-epoxy- and 9,10-dihydroxy saturated acids are related biosynthetically to the corresponding 9-monoenoic acids.

General Comments on Unusual Fatty Acids

(1) Tables 1–3 illustrate that a high proportion of the known unusual fatty acids have a C_{18} or, to a lesser extent, C_{16} chain length (cf. the major normal fatty acids).

(2) Most of the remaining unusual acids have C_{17} or C_{18+2n} carbon chains. In these, characteristic functional groups are at the same posi-

tions, relative to the methyl terminal (ω) end and can be related biochemically to the C_{18} homologs by either α-oxidation ($C_{18} \rightarrow C_{17}$) or by chain elongation ($C_{18} + nC_2 \rightarrow C_{18+2n}$) processes, both of which involve modifications at the carboxyl end of the chain; for this reason, the ω-numbering convention is used extensively in the tables and the following discussion.

(3) Substituent groups usually occur at α- or ω-terminal positions on the fatty acid chain or, internally, at positions which, in the common unsaturated fatty acid, are occupied by *cis*-ethylenic groups. The positions of substituent groups in unusual C_{18} acids are illustrated in Fig. 1.

Unusual Saturated Fatty Acids

Table 1 lists known oxygenated derivatives of saturated fatty acids. Other substituted saturated acids not included in the table include the cyclopropanoid acids which occur widely in plants of the family Malvaceae (see Smith, 1970; Yano *et al.*, 1972). Branched-chain acids, although common in microorganisms, are rare in higher plants; C_{16} and C_{17} branched acids have been reported in seeds of *Antirrhinum majus* (Radunz, 1965) and a C_{17} branched acid was recently demonstrated in the leaf lipids of several coniferous species (Jamieson and Reid, 1972).

The 2-D-hydroxy-saturated acids are characteristic components of cerebrosides and phytoglycolipids and occur widely in these forms; they have not been reported in seed triglycerides or in cutins. A range of ω- and (ω-1)-hydroxy fatty acids have been isolated from fruit was extracts (Mazliak, 1968).

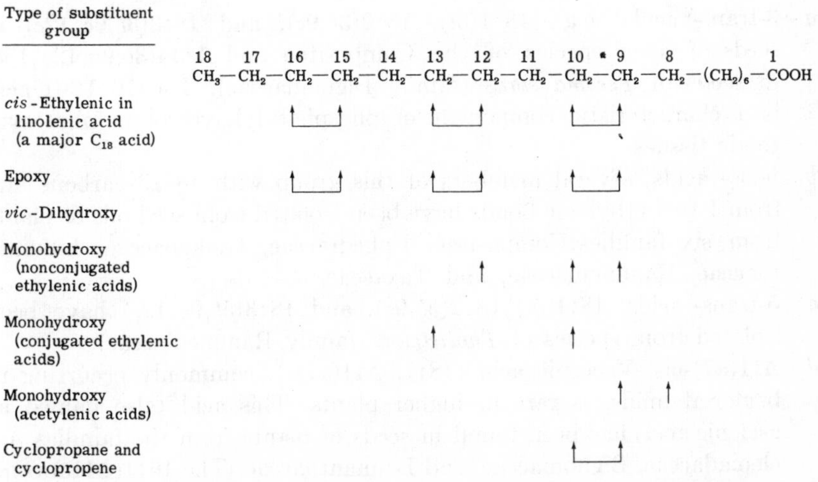

FIG. 1. Positions of in-chain substituent groups on unusual C_{18} acids in higher plants.

Hydroxyl groups at internal positions of saturated fatty acids are observed in cutin and novel members of this group, i.e., 9- and 10-hydroxy 14:0 acids were recently identified in leaf cutin from the coffee plant along with small amounts of their 7- and 8-hydroxy isomers (Holloway *et al.*, 1972). Various species of the Convolvulaceae contain glycosides of mono- and dihydroxy fatty acids in which the positions of the hydroxyl groups are so far unique in plant lipids.

Saturated fatty acids with either an epoxide ring or a vic-dihydroxy group in the 9,10-position occur in some seed oils (Table 1); the same structures with a C_{18} chain length and an additional hydroxy or oxo group at the $\omega1$-positions are characteristic components of cutins. The only fatty acid of this type from a noncutin source is the *threo*-9,10,18-trihydroxy 18:0 acid from the seed oil of *Champaepeuce afra*.

Unusual Nonconjugated Ethylenic Fatty Acids

The nonconjugated ethylenic acids from higher plants may be conveniently grouped on the basis of the position(s) and number of ethylenic centers. Table 2 shows the structural relationship between unusual ethylenic fatty acids and the commonly occurring fatty acids, oleic ($\omega9$), linoleic ($\omega6,9$), and linolenic ($\omega3,6,9$) acids.

Other classes of nonconjugated ethylenic fatty acids in higher plants are not included in Table 2 because, with one exception, oxygenated or other substituted forms of these acids are unknown. These classes, which are described in detail by Smith (1970) and Hitchcock and Nichols (1971), are as follows:

a   3-trans- acids; e.g., 18:1(3*t*), 18:2(3*t*,9*c*), and 18:3(3*t*,9*c*,12*c*) in seeds of some species of the Compositae and 18:4(3*t*,9*c*,12*c*,15*c*) in seeds of *Tecoma stans* (family Bignoniaceae). The 16:1(3*t*) acid is a characteristic component of phosphatidylglycerol in photosynthetic tissues.

b   5-cis- acids; several members of this group with 18–22 carbons and from 1 to 4 ethylenic bonds have been isolated from seed oils of plants from six families: Compositae, Ephedraceae, Ginkgoaceae, Limnanthaceae, Ranunculaceae, and Taxaceae.

c   5-trans- acids; 18:1(5*t*),18:2(5*t*,9*c*), and 18:3(5*t*,9*c*,12*c*) have been isolated from species of *Thalictrum* (family Ranunculaceae).

d $\Delta11(\omega7)$-cis. Vaccenic acid (18:1, $\Delta11(\omega7)c$), commonly occurring in bacterial lipids, is rare in higher plants. This acid (also named as asclepic acid) has been found in seeds of plants from the families Asclepiadaceae, Bignoniaceae, and Limnanthaceae. The 16:1($\Delta11c$) compound has been identified in a member of the Proteaceae and the homologous 20:1($\omega7c$) in rapeseed oil (Cruciferae).

The only substituted acids in any of the above four classes so far known are the oxygenated 3-trans acids in seed oil from a *Stenachaenium* sp.; Kleiman *et al.* (1971) have isolated from these seeds *cis*-9,10-epoxy-18:2(3t,12c) and two conjugated dienol derivatives in addition to the nonoxygenated 18:3(3t,9c,12c) acid.

With reference to Table 2, a series of ω9-monoenoic acids related to oleic acid occurs in the seed oils of certain plants from the families Cruciferae, Sapindaceae, Euphorbiaceae, and Compositae. The nonoxygenated acids include erucic acid (22:1, ω9c) which occurs in seeds of three-quarters of all Crucifeae species. Monohydroxy acids in this series include ricinoleic acid, which comprises over 90 percent of the seed triglyceride fatty acids in the castor bean *Ricinus* spp. (family Euphorbiaceae). Ricinoleic acid contains the hydroxyl group at the Δ12(ω7) position, and related acids, Δ12-hydroxy-16:1(ω9c) and ω7-hydroxy-20:1(ω9c) occur in seed oils of *Lesquerella* spp. (Cruciferae). The ω1-hydroxy acid in this series is a component of cutin. The 12,13-epoxyoleic acid, (+)-vernolic acid (*cis*-12-D, 13-D-epoxy-18:1, 9c) is the major (72 percent of total) fatty acid in *Vernonia anthelmintica* seeds but occurs in plants from at least five families. The levorotatory enantiomorph is found in some species of the family Malvaceae.

In the linoleic acid group (ω6,9-dienes) the 15, 16-epoxy-C_{18} derivative occurs in *Camelina sativa* seeds and the ω1-hydroxy-18:2(ω6,9) is a component of cutin in some plants, although polyunsaturated acids are not common constituents of cutin (Kolattukudy and Walton, 1972).

The only oxygenated acid in the linolenic acid group (ω3,6,9-trienes) is 2-D-hydroxylinolenic acid found in *Thymus vulgaris* seed oil.

Δ 6-cis Acids also are found. Petroselenic acid (18:1,6c) is a characteristic fatty acid in seed oils of plants (order Umbellales) of the families Umbelliferae and Araliaceae. The related trienoic γ-linolenic acid (18:3, 6c,9c,12c) and tetraenoic acid, stearidonic acid (18:4,6c,9c,12c,15c) occur widely in seeds of plants in the Boraginaceae family. Jamieson and Reid (1969, 1971a) have recently shown that γ-linolenic and stearidonic acids occur in relatively high proportions in the leaf lipids of members of the families Boraginaceae and Caryophyllaceae and that these acids are concentrated in somatic galactolipids.

Conjugated Ethylenic and Acetylenic Fatty Acids

In Table 3 the conjugated and nonconjugated acetylenic acids are listed with conjugated ethylenic acids. All known examples of these types have 18 carbon atoms or are presumed to be derived from C_{18} homologs by α-oxidation (C_{17} acids) or by chain cleavage (C_{10} and C_{12} acids); thus the ω-numbering convention again facilitates comparisons. All acetylenic and conjugated ethylenic fatty acids in higher plants have an unsaturated center at either the ω6 or ω9 position. With the exception of α-kamlolenic acid with its ω1-hydroxyl group, the hydroxyl group of acids in these classes occurs on the carbon adjacent to a trans-ethylenic or to an acetylenic bond to form conjugated dienol, enynol, diynol systems or systems with multiple conjugation.

In the ω6 unsaturated group (Table 3), the three known oxygenated derivatives, all found in plants from the Compositae family, have a D-hydroxyl group at the ω10-position and a trans-ethylenic bond at the ω8-position; they differ only in the configuration of the ω6 bond, i.e., either 6c or 6t (in dimorphecolic acid) or 6a (in helenynolic acid). Coexisting with dimorphocolic acid in seeds of *Dimorphotheca sinuata* is its corresponding oxo acid.

Keto derivatives of fatty acids occur relatively rarely in higher plants, but two other oxo acids with conjugated unsaturation occur in the ω9-unsaturated group (Table 3); these have a keto group in the Δ4(ω15) position and have been isolated from two species of the Rosaceae family. The most well established example of a conjugated dienol fatty acid in the ω9-group is coriolic acid from seeds of *Coriaria nepalensis*; the hydroxyl group at the ω6(Δ13) position has the D configuration, in common with almost all naturally occurring hydroxy fatty acids. However, Phillips *et al.* (1970) and Phillips and Smith (1972) have recently observed that in the seed oil of *Monnina emarginata*, (S)-coriolic acid (13-L-hydroxy-18:2,9c,11t) represents 30 percent of the total triglyceride fatty acid and is accompanied by a small amount of the corresponding 13-oxo derivative.

Isomeric ω6- and ω10-hydroxy-conjugated dienol fatty acids have been isolated from the same seed in several cases (Smith, 1970) and Kleiman *et al.* (1971) recently found the isomeric Δ3-trans-type acids, 9-hydroxy-18:3(3t,10t,12c), and 13-hydroxy-18:3(3t,9c,11t) in *Stenachaenium* sp. seed oil.

It is well established that autoxidation of polyunsaturated fatty acids leads to mixtures of ω6- and ω10-hydroperoxy acids, which on reduction can give corresponding dienol derivatives. Kleiman *et al.* (1971) found higher concentrations of the isomeric dienols in stored oils and these oxygenated derivatives could have been formed as artefacts during storage of the oil. Furthermore, oxygenation of seed oil fatty acids can presumably also take place during storage of intact seeds, possibly due to enzymic activity (Spencer *et al.*, 1973). It should be noted that autoxidation produces approximately equal amounts of racemic ω6- and ω10-oxygenated derivatives, whereas enzyme, e.g., lipoxygenase (discussed later) activity would be expected to produce one enantiomorph and, normally, an unequal distribution of isomeric forms.

Other Naturally Occurring Oxygenated Fatty Acids in Higher Plants

Other unusual oxygenated fatty acids, not shown in Tables 1–3, include a conjugated acid with a hydroxymethyl branch isolated from leaves and branches of an Australian plant, *Eremophila opposititflora* (family My-

$$CH_3(CH_2)_6CH \overset{trans}{=} CH—C \overset{trans}{=} CH—CH \overset{trans}{=} CH—COOH$$
$$|$$
$$CH_2OH$$

oporaceae). A unique fatty acid with a furanoid ring structure was obtained from the seed oil of *Exocarpus cupressiformis* (family Santalaceae).

$$CH_3(CH_2)_5—C\underset{O}{\overset{HC—CH}{\diagdown\diagup}}C—(CH_2)_7COOH$$

A 2-D-hydroxy derivative of sterculic acid (a cyclopropenoid fatty acid) was isolated from seed oils of *Pachira insignis* and of *Bombacopsis glabra* (both in the family Bombacaceae). Seed oils of some *Sapium* spp. contain an unusual hydroxy allenic C_8 fatty acid.

Oxygenated forms of ω-dicarboxylic acids have been demonstrated in cutins and suberins from higher plants. Monohydroxyhexadecane-1,16-diolic acid was isolated from cutin of the coffee plant (Holloway *et al.*, 1972) and 9,10-epoxyoctadecane-1,18-dioic acid together with its 9,10-dihydroxy analog was characterized in suberins from several plants by Holloway and Deas (1973).

Some unique oxygenated fatty acids which have not been found in healthy plant tissues, but which are formed enzymically in disrupted tissues, are described later (p. 233).

Unusual Fatty Acids and Chemotaxonomy

In a paper presented to an audience of phytochemists, a discussion of phylogenetic and taxonomic aspects of the subject would be appropriate. However, in the field of unusual fatty acids, particularly the oxygenated acids, such discussion can only be of limited value. At present, meaningful attempts at classification can only be made with unusual fatty acids of seed oils. Hitchcock and Nichols (1971) have revised and updated the definitive classification of seed oils fatty acids of Hilditch and Williams (1964).

Sufficient data are not yet available on cutin fatty acids, and studies on somatic lipids are unlikely to yield much useful information. Even with seed lipids, Hitchcock and Nichols (1971) have pointed out the limitations of any attempt at chemotaxonomic classification; these include differences in fatty acid composition of a given species due to environmental and genetic variations. Furthermore, species that are closely related morphologically often have widely different seed oil composition.

The epoxy acid, (+)-vernolic acid, occurs in some plants from five families with little botanical relationship (Compositae, Euphorbiaceae, Onagraceae, Dipsacaeae, and Valerianaceae). Unconjugated ethylenic fatty acids with $\Delta12(\omega7)$-hydroxy groups have been found in species from Cruciferae and Euphorbiaceae families. In a survey of the seed oils of the genus *Lesquerella* (family Cruciferae) Mikolajczak *et al.* (1962) observed that all species contained either lesquerolic acid ($\omega7$-hydroxy-18:1,$\omega9$) or densipolic acid ($\omega7$-hydroxy-18:2,$\omega3$,9). The $\omega6$-unsaturated conjugated dienol fatty acids (Table 3) have been isolated from various species of the Compositae and the analogous acetylenic acid (helenynolic acid) occurs in *Helichrysum bracteatum*, also a member of the Compositae.

In the $\omega9$-unsaturated class of conjugated acids, the known oxygenated derivatives are found in different families. However, the $\omega9$-acetylenic acids (the stearolic acid type), both oxygenated ($\omega11$-hydroxy-) and unoxygenated, are found widely in plants belonging to the two families Santalaceae and Olacaceae (see Table 3).

In the Convolvulaceae unusual mono- and dihdroxy acids (Table 1) occur widely in species of this single family.

Among the nonoxygenated unusual fatty acids, the cyclopentene fatty acids are known only in seeds from the family Flacourtiaceae (Hitchcock and Nichols, 1971). Although the cyclopropanoid and cyclopropenoid acids are confined to seeds of four families (Malvaceae, Sapindaceae, Sterculaceae, and Tiliaceae), the occurrence of these acids in somatic lipids of various plants (Kuiper and Stuiver, 1972) limits their taxonomic significance.

Biosynthesis of Oxygenated Fatty acids

Only the oxygenated unusual fatty acids are dealt with in this section. For information on biosynthesis of the nonoxygenated acids, the reader is referred to the recent review by Hitchcock and Nichols (1971). For a discussion of the biosynthetic mechanisms, it is convenient to divide the oxygenated acids into the following groups: (A) 2-hydroxy acids; (B) ω-oxygenated acids; (C) in-chain hydroxylated nonconjugated acids; (D) epoxy and vic-dihydroxy acids; and (E) oxygenated forms of fatty acids with conjugated unsaturation and derivatives formed in lipoxygenase-mediated enzyme reactions.

2-Hydroxy Acids

All naturally occurring 2-hydroxy acids from plants have the D-optical configuration. The literature contains references to two, apparently separate, plant enzyme pathways of α-oxidation leading to 2-D-hydroxy acids and, by further oxidation steps, to fatty acids with C_{n-1} chain lengths. Martin and Stumpf (1959) originally described a peroxidase system in peanut cotyledons requiring H_2O_2 for activity. Recently, Markovetz *et al.* (1972) demonstrated the formation of 2-D-hydroxy acids in this process. Hitchcock and James (1964, 1966) also showed the formation of 2-D-hydroxy acids in an enzyme system from pea leaves that required molecular oxygen for activity. In both these systems, concurrent formation of the 2-L-enantiomer was proposed as a transient intermediate in the further oxidative decarboxylation. Unpublished studies by W. Shine and P. K. Stumpf (to be reported at this meeting) indicate that the peanut cotyledon and the pea leaf systems are, in fact, the same. They suggest that the initial α-oxidation step is a free-radical process, involving molecular oxygen and a flavoprotein, forming a 2-D-hydroperoxy acid intermediate which can undergo subsequent reduction to the 2-D-hydroxy acid or decarboxylation to the C_{n-1} fatty acid. According to this hypothesis, the 2-L-hydroxy acid enantiomer is not involved at all in the α-oxidation process. Morris and Hitchcock (1968) have demonstrated that the formation of the 2-D-hydroxy acid involves direct attack by hydroxyl with retention of configuration at the 2-position and their results exclude the possibility of a dehydrogenation–hydration mechanism. The hydroxylation takes place with the free fatty acid as substrate and the enzyme(s) are soluble (Martin and Stumpf, 1959).

ω-Hydroxy and ω-Dicarboxylic Fatty Acids

A plant biochemist is always reluctant to preface a discussion with the words "although the mechanism in plants is not known, the following reaction has been established in rat liver or *E. coli*...". However, the only information on ω-oxidation in higher plants is based on incorporation studies of radioactive substrates into the tissues of leaf and fruit skin. Kolattukudy (1970, 1972) has studied the incorporation of ^{14}C-labeled palmitic acid into 16-hydroxpalmitic acid leaf disks and apple fruit peel. Oxygen was required for the incorporation which was blocked by the iron chelator, 1,10-phenanthroline. These data indicate that a mixed-function oxidase reaction is probably operative as in ω-oxidation systems in animals and microorganisms. All such systems require molecular oxygen, a nonheme-iron protein, and a hydrogen donor (usually NADH or NADPH). In mammalian systems (Lu *et al.*, 1969) and in the yeast,

Candida utilis (Lebault *et al.*, 1971), cytochrome P-450 is involved. In *Pseudomonas* sp. cytochrome P-450 is not involved (McKenna and Coon, 1970).

The formation of a dicarboxylic acid from the corresponding ω-1-hydroxy acid has been demonstrated in a cell-free system from epidermal tissue of *Vicia faba* leaves. Kolattukudy (1972) obtained NADP-specific, ω-1-hydroxy acid dehydrogenase activity in a particle-free supernatant preparation from this tissue.

IN-CHAIN MONOHYDROXY NONCONJUGATED FATTY ACIDS

The location of monohydroxy groups on the same carbon atoms as those carrying ethylenic bonds in the common (oleic, linoleic, and linolenic) acids, formally suggests a hydration mechanism for their formation. However, in the case studied in detail with higher plants, this is demonstrably not so.

Considerable attention has been paid to the biosynthesis of ricinoleic acid (12-D-hydroxy-18:1,9c), the major component of the seed triglyceride fatty acids of castor bean. Ricinoleic acid also occurs in ergot oil from the fungus *Claviceps purpurea* and is formed in this organism by hydration of linoleic acid (Morris *et al.*, 1966). However, in developing castor bean seeds, linoleic acid is not a precursor of ricinoleic acid (James *et al.*, 1965; Galliard and Stumpf, 1966). Using a microsomal fraction from immature castor bean seeds Galliard and Stumpf (1966) were able to demonstrate that oleyl-CoA was sterospecifically hydroxylated by a mixed-function oxidase type of reaction in which molecular oxygen, NADH, and an iron-containing protein were required. The enzyme reaction was highly specific for the oleyl-CoA substrate (18:1,9c); no hydroxylation was obtained with the CoA-thioesters of stearic acid (18:0), linoleic acid (18:2,9c,12c), elaidic acid (18:1,9t), or *cis*-vaccenic acid (18:1,11c). The sterospecificity of the reaction was confirmed by Morris (1967), who showed that the hydroxylation at the 12-D-position of oleic acid proceeded with retention of configuration.

The 10, 16-dihydroxypalmitic acid of cutin is also formed by direct hydroxylation of the saturated 16-hydroxypalmitoyl-CoA. In a cell-free system from *Vicia faba* epidermal tissue, CoASH, ATP, and NADPH were required for the formation of the conversion of the ω1-hydroxy acid to the dihydroxy derivative (Walton and Kolattukudy, 1972). Thus, as with ricinoleic acid biosynthesis, activation of the carboxyl group as a CoA-thioester was a prerequisite for reaction and mixed-function oxidase reaction was probably involved. However, the occurrence of the in-chain hydroxyl group of cutin fatty acids at more than one position on the chain (see Table 1) suggests that the hydroxylation system may lack absolute positional specificity.

Radioactive tracer studies in which ^{14}C-labeled stearic and oleic acids were administered to cutin tissue led Kolattukudy and Walton (1972) to suggest that the in-chain hydroxyl groups of 10,18-dihydroxystearic acid could be formed, not by direct hydroxylation, but either by reduction of a 9,10-epoxide or by hydration of the ethylenic bond in 18-hydroxyoleic acid.

Epoxy and vic-Dihydroxy Fatty Acids

In the absence of data on cell-free systems, several alternative theories have been proposed for the biosynthesis of epoxy and vic-dihydroxy acids. A proposal by Miwa *et al.* (1963) that vic-dihydroxy acids are precursors of epoxy acids is not substantiated by more recent work, which, in general supports the reverse conclusion that epoxy acids are hydrated to dihydroxy derivatives. In mammalian cells epoxides have been shown to be obligatory intermediates in the conversion of olefins to vic-dihydroxy acids and the initial oxygenation of the olefin is catalyzed by a mixed-function oxidase system (Maynert *et al.*, 1970). A similar process is involved in steroid and vic-diol formation (Oesch and Daley, 1971).

Based on results from incorporation of labeled substrates into plant tissues and on the observed occurrence together of epoxy and vic-dihydroxy acids in cutin, several workers have recently given support to the operation of the following pathway in higher plants (Conacher and Gunstone, 1969; Morris, 1970; Kolattukudy and Walton, 1972; Holloway and Deas, 1973):

$$-CH\!\!=\!\!CH- \xrightarrow[\text{[H]}]{O_2} \overset{O}{\overset{/\ \backslash}{-CH-CH-}} \xrightarrow{H_2O} -CHOH-CHOH-$$
$$\text{\small cis}$$

The epoxidation reaction in cutin and suberin formation appear to be specific for the Δ9-position of C_{18} olefinic fatty acids (Holloway and Deas, 1973). The stereospecificity of the enzymatic hydration of (+)-vernolic acid (12-D,13-D-epoxy 18:1,9c) was studied in seeds of *Vernonia anthelmintica* by Morris and Crouchman (1969), who showed that hydroxyl attack at the 12-position produced inversion at that carbon and the formation of *threo*-12-L, 13-D-dihydroxy 18:1(9c).

An alternative theory for the biosynthesis of epoxy and polyhydroxy fatty acids in cutin and suberin and also in seeds involves the participation of the enzyme, lipoxygenase. Heinen and van den Brand (1963) showed that the activity of this enzyme was increased in damaged leaf tissues during suberization. Recent work in our laboratory (M. J. Oliver and T. Galliard, unpublished observations) has shown a similar increase in lipoxygenase activity in wounded tissue from potato tubers, but the area of maximal lipoxygenase increase was a considerable distance away from the area of suberin formation. Recently, Spencer *et al.* (1973) demonstrated that in seeds of *Crepis vesicaria* during storage, parallel increases were observed in the enzymatic formation of conjugated dienoic fatty acids and of epoxy acids. These authors suggested that lipoxygenase was involved in the formation of both types of oxygenated acids and they proposed that perepoxide derivatives were intermediates in epoxide formation.

Although the weight of evidence supports the epoxidation hydration pathway, lipoxygenase may be involved in the biopolymerization reaction of cutin and suberin synthesis, particularly in the formation of peroxide bonds (Kolattukudy and Walton, 1972).

LIPOXYGENASE-MEDIATED REACTIONS AND FATTY ACIDS WITH CONJUGATED UNSATURATION

In the absence of definitive work on the biosynthesis of fatty acids with conjugated unsaturation and because of their formal resemblance to the products of lipoxygenase-mediated reactions, discussion of these topics is combined.

Lipoxygenase

Although the enzyme lipoxygenase (EC 1.13.1.13) has been known for many years and it was one of the first enzymes obtained in a crystalline form (Theorell *et al.*, 1947), its physiological role remains enigmatic. The occurrence of the enzyme was originally thought to be confined to seeds of leguminous plants and some cereals (Tappel, 1963), but it is now known to be more widely distributed among higher plants (Pinsky *et al.*, 1971; Grosch, 1972) occurring also in leaves (Holden, 1970), A particularly rich source is the tuber potato, *Solanum tuberosum* (Galliard, 1970).

Lipoxygenases isolated from different sources differ in several respects, including substrate specificity, pH optima, and, most significantly, isomeric forms of products formed. Even from a single source, soy beans, several isoenzymic forms with different properties have been isolated (Christopher *et al.*, 1970).

Basically, all lipoxygenase enzymes acting under aerobic conditions catalyze the hydroperoxidation of long-chain fatty acids containing an $\omega 6c,9c$-diene structure to form $\omega 6$-hydroperoxy-$\omega 7t,9c$-diene and/or $\omega 10$-hydroperoxy-$\omega 6c,8t$-diene derivatives; the positional specificity is to the methyl terminal portion of the fatty acid chain (Hamberg and Samuelsson, 1965)

$$—CH{=}CH—CH_2—CH{=}CH \quad \rightarrow$$
$$\quad \textit{cis} \qquad\qquad\qquad \textit{cis}$$

$$—CH{=}CH—CH{=}CH—CH(OOH)— \quad \textit{or} \quad —CH(OOH)—CH{=}CH—CH{=}CH—$$
$$\quad \textit{cis} \qquad \textit{trans} \qquad\qquad\qquad\qquad\qquad \textit{trans} \qquad \textit{cis}$$

The hydroperoxides formed under the action of lipoxygenase are optically active. For example, linoleic acid is converted to (-)-9-D-hydroperoxy-18:2 (9c, 11t) and (+)-13-L-hydroperoxy-18:2(10t,12c). One isoenzyme from soybean (Christopher and Axelrod, 1971) forms mainly the 13-L-isomer, the lipoxygenases from corn (Gardner and Weisleger, 1970) and potato (Galliard and Phillips, 1971) form almost exclusively the 9-D-isomer, and other lipoxygenases appear to produce mixtures of the two. The relative amounts of the two isomeric hydroperoxides produced by the enzyme from some tissues depend, to some extent, on oxygen tension, pH, and temperature (Christopher *et al.*, 1972).

Hamberg and Samuelsson (1967) and Egmond *et al.* (1972) have shown that the initial reaction involves the stereospecific abstraction of a hydrogen atom from the ω8-methylene group of the substrate; the removal of the D_R-hydrogen at this position leads to the formation of the ω10-D-hydroperoxide isomer, wheras formation of the ω6-L-hydroperoxide isomer involves removal of the L_S-hydrogen at the ω8-carbon.

Until very recently, lipoxygenase was thought to be devoid of prosthetic groups or metals (see Tappel, 1963), but independant studies by Chan (1972, 1974) and Roza and Franke (1974) have shown that soybean lipoxygenase contains one atom of nonheme iron per mole of enzyme (molecular weight *c.* 10^5 daltons).

Further Enzymatic Reactions of Fatty Acid Hydroperoxides

A fatty acid hydroperoxide isomerase has been studied in cell-free extracts of flax seed, *Linum usitatissimum* (Zimmerman, 1966; Zimmerman and Vick, 1970) and corn, *Zea mais* (Gardner, 1970). The flax enzyme converts the 13-L-hydroperoxy-18: 2(9c, 11t) into an α-ketol derivative:

$$CH_3—(CH_2)_4—CH(OOH)—CH{=}CH—CH{=}CH—(CH_2)_7COOH\rightarrow$$

$$CH_3—(CH_2)_4—CHOH—CO—CH_2—CH{=}CH—(CH_2)_7COOH$$

The 9-D-hydroperoxide isomer is not attacked by the flax seed enzyme (Veldink *et al.*, 1970a). In corn extracts, linoleic acid is converted *via* hydroperoxide intermediates to 9-hydroxy-10-oxo-18:1(12c) acid and the isomeric 13-hydroxy-10-oxo-18:1(10t) acid. The corn enzyme also produced 13-hydroxy-12-oxo-18:1(9c)- and 9-hydroxy-12-oxo-18:1(10t) acids when incubated with mixed 9-D- and 13-L-hydroperoxides of linoleic acid. Veldink *et al.* (1970b) made the interesting observation that in the 12-

oxo-13-hydroxy-18:1(9c) acid formed from 13-hydroperoxy-18:2(9c,11t) by the flax isomerase enzyme, only the carbonyl oxygen atom at position-12 came from the original hydroperoxy group at position-13 of the substrate, and these authors suggested a cyclic peroxide or an epoxide intermediate in the isomerization reaction.

Enzymes from potato tuber metabolize fatty acid hydroperoxides by a different route. Galliard and Phillips (1972) showed that a significant proportion of the total lipid from tubers extracted in aqueous media at pH 6 to 8 was represented by novel butadienylvinyl ether derivatives of linoleic and linolenic acids. These workers demonstrated the enzymatic formation of the ether derivatives via 9-D-hydroperoxide intermediates, e.g., for linoleic acid

$$CH_3—(CH_2)_4—\underset{cis}{CH=CH}—CH_2—\underset{cis}{CH=CH}—CH_2—(CH_2)_6COOH \xrightarrow{\text{lipoxygenase}}$$

$$CH_3—(CH_2)_4—\underset{cis}{CH=CH}—\underset{trans}{CH=CH}—CH(OOH)—CH_2—(CH_2)_6COOH \xrightarrow{\text{enzyme}}$$

$$CH_3—(CH_2)_4—\underset{cis}{CH=CH}—\underset{trans}{CH=CH}—O—\underset{trans}{CH=CH}—(CH_2)_6COOH$$

An analogous ether derivative with an additional *cis*-ethylenic group at the ω3-position is formed from linolenic acid (Galliard *et al.*, 1973).

The lipoxygenase from potato formed only the 9-D-hydroperoxide from linoleic acid and the 13-L-isomer was inactive as a substrate for the subsequent formation of the ether derivative. The reaction is of particular interest because it involves the insertion of an atom of oxygen into a fatty acid chain. Subsequent studies (Galliard *et al.*, 1974) have shown that an enzyme in potato tubers catalyzes the subsequent breakdown of the derived ether to volatile carbonyl fragmentation products. There is some evidence that this sequence of at least three reactions may all be catalyzed by the lipoxygenase enzyme (unpublished observations). The above reactions are part of a sequence of enzymatic reactions that are initiated by disruption of tuber tissue; the first reaction of the sequence is the liberation of linoleic and linolenic acids from membrane-bound phospholipids and galactolipids by a lipolytic acyl hydrolase (Galliard, 1970, 1971a,b). This chain of hydrolytic and oxidative enzyme reactions is of significance in the degradative processes that occur with membrane structures in aqueous extracts of plant tissues and relates to the formation of volatile flavor (and off-flavor) products formed by disruption of some plant tissues.

A range of degradation products are formed under anaerobic conditions with soybean lipoxygenase, linoleic acid, and its 13-L-hydroperoxide (Garssen *et al.*, 1971, 1972). These include 13-oxo-18:2(9c,11t) and the fragmentation products 13-oxo-13:2(9*e*,11*t*) and pentane, in addition to oxygenated and nonoxygenated dimeric fatty acid derivatives. No products were formed in this system when the 13-L-hydroperoxide was replaced by the 9-D-isomer. These workers have speculated that the lipoxygenase-mediated processes may play a physiological role in seeds by maintaining a low oxygen tension (Garssen *et al.*, 1971).

Lipoxygenase, a hydroperoxide isomerase, and endogenous proteins have been implicated in the formation of a range of hydroperoxy and hydroxy diene acids together with epoxy and polyhydroxy derivatives in aqueous suspensions of wheat flour (Graveland, 1970a,b) and barley (Graveland *et al.*, 1972).

Possible Role of Lipoxygenase in the Biosynthesis of Fatty Acids with Ethylenic Conjugation

Gunstone (1966) proposed that the known conjugated dienol fatty acids could be formed from linoleic acid via 9,10- and 12,13-epoxy intermediates but to date no experimental evidence has implicated free epoxide intermediates via this process.

An alternative proposed by Morris and Marshall (1966) involved a lipoxygenase type of reaction in which, for example, linoleic acid (18:2, 9*c*,12*c*) could be converted to 9-hydroxy-18:2(10*t*,12*c*) and 13-hydroxy-18:2(9*c*,11*t*) (coriolic acid)—both naturally occurring fatty acids (see Table 3). Subsequent dehydration (which readily occurs by chemical reactions) would produce known conjugated trienoicacids (Table 3). A similar mechanism could be involved to explain the biosynthesis of oxygenated acetylenic acids with conjugated unsaturation.

However, as yet there is no evidence from biochemical studies for lipoxygenase involvement in the formation of naturally occurring conjugated ethylenic or acetylenic fatty acids. The formal resemblance between the products formed from linoleic and linolenic acids by lipoxygenase activity (i.e., the ω6-hydroperoxy-ω7,9-diene and ω10-hydroperoxy-ω6,8-diene structures) and the corresponding hydroxydiene structures in known conjugated acids become less direct in many cases when the absolute configuration and ethylenic geometry of these compounds is considered.

For example, the ω6(Δ13)-hydroperoxide group in lipoxygenase-mediated products has the L-absolute configuration, whereas coriolic acid (13-D-hydroxy-18:2,9*c*,11*t*) in seeds of *Coriaria nepalensis* has the opposite absolute configuration. However, the recently characterized Ls-coriolic acid (13-L-hydroxy-18:2,9*c*,11*t*) from *Monnina emarginata* seed oil (Phillips *et al.*, 1970) does have the same absolute configuration as the reduction product of the lipoxygenase-mediated product and could be an interesting subject for biosynthetic studies. Dimorphecolic acid (9-D-hydroxy-18:2, 10*t*,12*t*), which represents two-thirds of the total fatty acid content in seeds of *Dimorphotheca sinuata*, differs from the reduction product of lipoxygenase-mediated 9-D-hydroperoxy-18:2(10*t*,12*c*) in the configuration of the ethylenic bond at the 12-position. Gardner *et al.* (1973) recently studied the possible role of lipoxygenase in this tissue. Although lipoxygenase was present in the seed, the enzyme showed specificity for the 13-L-hydroperoxidation, forming the usual cis, trans-diene structure of lipoxygenase-catalyzed reactions. The authors pointed out that the studies were performed with extracts from mature seeds and that a different type of lipoxygenase could be involved during the period of seed oil biosynthesis in developing seeds. It is well established that some specific enzymes of seed oil biosynthesis are only demonstrated during a limited period of development (see below).

Thus the role of lipoxygenase in the biosynthesis of conjugated dienol fatty acids has yet to be established. Although known lipoxygenase reactions followed by reduction of the hydroperoxides to corresponding hydroxy derivates are compatible with the structures of some naturally occurring fatty acids, e.g., 9-D-hydroxy-18:2(10*t*,12*c*) in seeds of *Calendula officinalis* and *Xeranthemum annum* (see Table 3) and L-coriolic acid mentioned above, some major modifications to the established enzyme reactions would be necessary to explain the formation of D-coriolic acid, dimorphecolic acid, and some of the nonoxygenated conjugated acids listed in Table 3.

Little is known about the biosynthesis of hydroxylated acetylenic fatty acids. Current knowledge on naturally occurring acetylenic compounds has been reviewed by Bohlmann *et al.* (1973) and biosynthetic aspects of acetylenic fatty acids are discussed by Bu'Lock and Smith (1967), Haigh *et al.* (1969), and Hitchcock and Nichols (1971). The biosynthetic pathway for helenynolic acid (9-D-hydroxy-18:2, 10*t*,12*a*) could be analogous to that for dimorphecolic acid (9-D-hydroxy-18:2, 10*t*,12*t*). The acetylenic fatty acids of the stearolic acid group (ω9*a*) all have the hydroxyl group at the ω11-position (see Table 3), and Hitchcock and Nichols (1971) have suggested that hydroxylation occurs at the ω11(Δ8) methylene group of the corresponding C_{18} acid and that the C_{17} homologs are subsequently formed by α-oxidation.

Chain Length Modifications

As mentioned earlier, the majority of unusual fatty acids are either C_{18} acids or are C_{17} or C_{18+2n} acids that can be formally related to homologous C_{18} acids by α-oxidation (-C1) or chain elongation ($+ nC_2$).

The formation of n-1 fatty acids by α-oxidation is well established *in vivo* and *in vitro* with higher plants (see Hitchcock and Nichols, 1971).

Chain elongation processes are also well established (Hawke and Stumpf, 1965) and the elongation of ricinoleic acid (ω7-hydroxy-18:1, ω9c) to lesquerolic acid (ω7-hydroxy-20:1, ω9c) was demonstrated by Yang and Stumpf (1965), who obtained a mitochrondrial preparation from avocado mesocarp tissue that catalyzed conversion of ricinoleic acid to the C_{20} homolog in the presence of acetyl CoA as the C_2 donor.

Regulatory Factors and Unusual Fatty Acids

The aspects discussed below are relevant to the occurrence of unusual fatty acids in plants and to developmental aspects of their metabolism.

ENVIRONMENTAL ASPECTS

The degree of unsaturation in oil lipids of seeds from plants in their natural environment is well correlated with climatic ambient temperature (Hilditch and Williams, 1964). This can be generalized in that the higher the average ambient temperature, the higher the proportion of more saturated fatty acids and a resulting higher melting point for the oils. A similar relationship has been found for the membrane lipids of alfalfa grown at low and high temperatures (Kuiper, 1970), and this effect is associated with relative cold-hardiness of plant tissues (Lyons and Raison, 1970).

Incorporation of oxygenated groups into fatty acid chains has the same effect as increasing the degree of saturation in that melting points are increased. Thus at a given temperature lipids containing oxygenated acids are less fluid than corresponding nonoxygenated forms. Mazliak (1968) has summarized the evidence showing that leaf and stem waxes from tropical plants have longer chain length acids, less unsaturation, and more oxygenated acids than waxes from temperate plants; waxes from subtropical plants showed intermediate properties. Correlations with flower and fruit waxes were less significant.

Canvin (1965) demonstrated that the composition of seed oils of given varieties of rape, sunflower, and flax depended on growing temperature in controlled experiments; however, safflower oil and castor bean oil (90 percent of total fatty acid is ricinoleic acid) were not affected by temperature differences.

Harris and James (1969) have proposed that temperature effects can be related to the concentration of tissue oxygen available to enzymes catalyzing desaturation reactions and presumably this concept could be extended to include oxygenation reactions.

Structural and Genetic Factors

In plants, as in most organisms, fatty acids with 16 and 18 carbon atoms are predominant in lipids with widely different functional roles, i.e., as storage lipids, membrane components, or surface lipids. Even for fatty acids with other chain lengths, the basic level for biochemical modifications of structure is the C_{16} or C_{18} precursor. Thus the *de novo* fatty acid synthetases in plant tissues are mainly programmed to produce these essential components.

Considerable variation may occur in the fatty acid composition of oil seeds from a given variety of plant grown under identical conditions. Plant breeders are making use of this genetic variation in work on commercially important crops. For example, strains of rapeseed exist that contain no erucic acid (normally up to 50 percent of the fatty acids) and it is known that a single gene controls the elongation of oleic (18:1, $\omega 9c$) to erucic (22:1, $\omega 9c$) acid (Appelqvist, 1969).

Developmental Aspects

Chemical analyses of the lipid composition of tissues give data on the net result of metabolism over the life of that tissue and may give rise to misleading conclusions on the biochemistry of the tissue at the time of analysis. As examples, one can quote the post-harvest changes in the oxygenated fatty acids of seed lipids (mentioned earlier) and the transient appearance of enzyme activities involved in the biosynthesis or breakdown of lipids. The mature castor bean does not contain activity for conversion of oleic acid to ricinoleic acid; the enzyme activity can only be demonstrated and isolated during a limited period of development in the seed (Yamada and Stumpf, 1964) which corresponds to the time during which ricinoleic acid is being produced *in vivo* (Canvin, 1963). Presumably, synthesis or activation of the enzyme(s) is subject to strict regulation. An analogous situation exists in germinating oil seeds where glyoxylate bypass enzymes involved in lipid-to-sugar conversion are maximally active over a relatively short time period (Beevers, 1961).

Control at the Cellular Level

The occurrence in many seeds of unusual fatty acids in the triglycerides and the absence of these in somatic lipids indicates a control either by enzyme–substrate compartmentalization or by product specificity in biosynthesis. Little is known about the regulatory mechanisms involved. Biosynthesis of the characteristic fatty acids of cutin is localized in cells of epidermal tissue (Kolattukudy and Walton, 1972) and specialized systems must operate in these cells to isolate the cutin components from other cellular lipid systems and to transfer these to the cuticular surface.

The subcellular localization of enzymes responsible for the biosynthesis and catabolism of oxygenated fatty acids is a regulatory factor depending on transfer of substrates, intermediates, and cofactors between cellular organelles. Interest in this area is now increasing, particularly in relation to developmental aspects.

The formation of unusual fatty acids in disrupted tissues demonstrates the reverse process, i.e., the loss of control at the cellular level. Cell disruption by damage, infection, and senescence leads to enzymatic reactions, which in healthy tissues must be either inactivated or prevented by spatial separation of substrates and enzymes. The sequence of hydrolytic and oxidative reactions that is initiated in disrupted tuber tissues from potato (see p. 234) illustrates this point. Both the lipolytic acyl hydrolase enzyme and lipoxygenase (together with the enzymes of subsequent oxidative sequence) are obtained as "soluble" enzymes even when the mildest techniques for organelle preparation are used. That these enzymes are unlikely to be present *in vivo* in the cytosol is shown by the fact that their activities *in vitro* are sufficient to destroy all the endogenous membrane lipids of the tissue in milliseconds if the substrates were available to the enzymes *in vivo*. The autodegradative nature of these processes limits studies on subcellular localization because the organelles themselves are attacked during isolation. Using tissues with less suicidal levels of these enzymes, we found that the hydrolytic enzyme at least is concentrated in a particulate fraction (D. A. Wardale and T. Galliard, unpublished).

ACKNOWLEDGMENTS

I am grateful to the program organizers and to the Phytochemical Society of North America for their kind invitation to prepare this review.

REFERENCES

Appelqvist, L.-A. 1969. *Hereditas* **61**:9.
Beevers, H. 1961. *Nature (London)* **191**:433.
Bohlmann, F., T. Burkhardt, and C. Zdero. 1973. "Naturally Occurring Acetylenes." Academic Press, New York.
Bu'Lock, J. D., and G. N. Smith. 1967. *J. Chem. Soc., C* p. 332.
Canvin, D. T. 1963. *Can. J. Biochem. Physiol.* **41**:1879.
Canvin, D. T. 1965. *Can. J. Bot.* **43**:63.
Chan, H.W.-S. 1972. *Commun. World Congr. Int. Soc. Fat Res., 11th, 1972.*

Chan, H.W.-S. 1973. *Biochim. Biophys. Acta* **327**:32.

Christopher, J., and B. Axelrod. 1971. *Biochem. Biophys. Res. Commun.* **44**:731.

Christopher, J., E. Pistorius, and B. Axelrod. 1970. *Biochim. Biophys. Acta* **198**:12.

Christopher, J. P., E. K. Pistorius, F. E. Regnier, and B. Axelrod. 1972. *Biochim. Biophys. Acta* **289**:82.

Conacher, H. B. S., and F. D. Gunstone. 1969. *Chem. Phys. Lipids* **3**:191.

Egmond, M. R., J. F. G. Vliegenthart, and J. Boldingh. 1972. *Biochem. Biophys. Res. Commun.* **48**:1055.

Galliard, T. 1970. *Phytochemistry* **9**:1725.

Galliard, T. 1971a. *Biochem. J.* **121**:379.

Galliard, T. 1971b. *Eur. J. Biochem.* **12**:90.

Galliard, T., and D. R. Phillips. 1971. *Biochem. J.* **124**:431.

Galliard, T., and D. R. Phillips. 1972. *Biochem. J.* **129**:743.

Galliard, T., and P. K. Stumpf. 1966. *J. Biol. Chem.* **241**:5806.

Galliard, T., D. R. Phillips, and D. J. Frost. 1973. *Chem. Phys. Lipids* **11**:173.

Galliard, T., D. A. Wardale, and J. A. Matthew. 1974. *Biochem. J.* **138**:23.

Gardner, H. W. 1970. *J. Lipid Res.* **11**:311.

Gardner, H. W., and D. Weisleder. 1970. *Lipids* **5**:678.

Gardner, H. W., D. D. Christianson, and R. Kleiman. 1973. *Lipids* **8**:271.

Garssen, G. J., J. F. G. Vliegenthart, and J. Boldingh. 1971. *Biochem. J.* **122**:327.

Garssen, G. J., J. F. G. Vliegenthart, and J. Boldingh. 1972. *Biochem. J.* **130**:435.

Graveland, A. 1970a. *J. Amer. Oil Chem. Soc.* **47**:352.

Graveland, A. 1970b. *Biochem. Biophys. Res. Commun.* **41**:427.

Graveland, A., L. Pesman, and P. van Erde. 1972. *Tech. Quart., Master Brew. Ass. Amer.* **9**:98.

Grosch, W. 1972. *Fette, Seifen, Anstrichm.* **74**:375.

Gunstone, F. D. 1966. *Chem. Ind. (London)* p. 1551.

Haigh, W. G., R. Safford, and A. T. James. 1969. *Biochim. Biophys. Acta* **176**:647.

Hamberg, M., and B. Samuelsson. 1965. *Biochem. Biophys. Res. Commun.* **21**:531.

Hamberg, M., and B. Samuelsson. 1967. *J. Biol. Chem.* **242**:5329.

Harris, P., and A. T. James. 1969. *Biochem. J.* **112**:325.

Hawke, J. C., and P. K. Stumpf. 1965. *Plant Physiol.* **40**:1023.

Heinen, W., and I. van den Brand. 1963. *Z. Naturforsch. B* **18**:67.

Hilditch, T. P., and P. N. Williams. 1964. "The Chemical Constitution of the Natural Fats," 4th ed. Chapman & Hall, London.

Hitchcock, C., and A. T. James. 1964. *J. Lipid Res.* **5**:593.

Hitchcock, C., and A. T. James. 1966. *Biochim. Biophys. Acta* **116**:413.

Hitchcock, C., and B. W. Nichols. 1971. "Plant Lipid Biochemistry." Academic Press, New York.

Holden, M. 1970. *Phytochemistry* **9**:507.

Holloway, P. J., and A. H. B. Deas. 1973. *Phytochemistry* **12**:1721.

Holloway, P. J., A. H. B. Deas, and A. M. Kabaara. 1972. *Phytochemistry* **11**:1443.

James, A. T., H. C. Hadaway, and J. P. W. Webb. 1965. *Biochem. J.* **95**:448.

Jamieson, G. R., and E. H. Reid. 1969. *Phytochemistry* **8**:1489.

Jamieson, G. R., and E. H. Reid. 1971a. *Phytochemistry* **10**:1575.

Jamieson, G. R., and E. H. Reid. 1971b. *Phytochemistry* **10**:1837.

Jamieson, G. R., and E. H. Reid. 1972. *Phytochemistry* **11**:269.

Kaimal, T. N. B., and G. Lakshminarayana. 1972. *Phytochemistry* **11**:1617.

Kleiman, R., G. R. Spencer, L. W. Tjarks, and F. R. Earle. 1971. *Lipids* **6**:617.

Kleiman, R., G. F. Spencer, F. R. Earle, H. J. Nieschlag, and A. S. Barclay. 1972. *Lipids* **7**:660.

Kolattukudy, P. E. 1970. *Biochem. Biophys. Res. Commun.* **41**:299.

Kolattukudy, P. E. 1972. *Biochem. Biophys. Res. Commun.* **49**:1040.

Kolattukudy, P. E. 1973. *Lipids* **8**:90.

Kolattukudy, P. E., and T. J. Walton. 1972. *Progr. Chem. Fats Other Lipids* **13**:121.

Kuiper, P. J. C. 1970. *Plant Physiol.* **45**:684.

Kuiper, P. J. C., and B. Stuiver. 1972. *Plant Physiol.* **49**:307.

Lebeault, J. M., E. T. Lode, and M. J. Coon. 1971. *Biochem. Biophys. Res. Commun.* **42**:413.

Lu, A. Y. H., K. W. Junk, and M. J. Coon. 1969. *J. Biol. Chem.* **244**:3714.

Lyons, J. M., and J. K. Raison. 1970. *Plant Physiol.* **45**:386.

McKenna, E. J., and M. J. Coon. 1970. *J. Biol. Chem.* **245**:3882.

Markovetz, A. J., and P. K. Stumpf. 1972. *Lipids* **7**:159.

Martin, R. O., K. Stumpf, and S. Hammarström. 1959. *J. Biol. Chem.* **234**:2548.

Maynert, E. W., R. L. Foreman, and T. Watabe. 1970. *J. Biol. Chem.* **245**:5234.

Mazliak, P. 1968. *Progr. Phytochem.* **1**:49–111.

Mikolajczak, K. L., F. R. Earle, and I. A. Wolff. 1962. *J. Amer. Oil. Chem. Soc.* **39**:78.

Miwa, T. K., F. R. Earle, G. C. Miwa, and I. A. Wolff. 1963. *J. Amer. Oil Chem. Soc.* **40**:225.

Morris, L. J. 1967. *Biochem. Biophys. Res. Commun.* **29**:311.

Morris, L. J. 1970. *Biochem. J.* **118**:681.

Morris, L. J., and M. L. Crouchman. 1969. *Lipids* **4**:50.

Morris, L. J., and C. Hitchcock. 1968. *Eur. J. Biochem.* **4**:146.

Morris, L. J., and M. O. Marshall. 1966. *Chem. Ind. (London)* p. 1493.

Morris, L. J., S. W. Hall, and A. T. James. 1966. *Biochem. J.* **100**:29c.

Oesch, F., and J. Daley. 1971. *Biochim. Biophys. Acta* **277**:692.

Phillips, B. E., and C. R. Smith. 1972. *Lipids* **7**:215.

Phillips, B. E., C. R. Smith, and L. W. Tjarks. 1970. *Biochim. Biophys. Acta* **210**:353.

Pinsky, A., S. Grossman, and M. Trop. 1971. *J. Food. Sci.* **36**:571.

Pohl, P., and H. Wagner. 1972a. *Fette, Seifen, Anstrichm.* **74**:424.

Pohl, P., and H. Wagner. 1972b. *Fette, Seifen, Anstrichm.* **74**:541.

Radunz, A. 1965. *Hoppe-Seyler's Z. Physiol. Chem.* **341**:192.

Roza, M., and A. Franke. 1973. *Biochim. Biophys. Acta* **327**:24.

Smith, C. R. 1970. *Progr. Chem. Fats Other Lipids* **11**:139–177.

Spencer, G. F., F. R. Earle, I. A. Wolff, and W. H. Tallent. 1973. *Chem. Phys. Lipids* **10**:91.

Tappel, A. L. 1963. *In* "The Enzymes" (P. D. Boyer, H. Lardy, and K. Myrbäck, eds.), 2nd ed., Vol. 8, pp. 275–283. Academic Press, New York.

Theorell, H., R. T. Holman, and Å. Åkeson. 1947. *Acta Chem. Scand.* **1**:571.

Veldink, G. A., J. F. G. Vliegenthart, and J. Boldingh. 1970a. *Biochem. J.* **120**:55.

Veldink, G. A., J. F. G. Vliegenthart, and J. Boldingh. 1970b. *Febs. Lett.* **7**:188.

Vijayalakshmi, B., and S. V. Rao. 1972. *Chem. Phys. Lipids* **9**:82.

Walton, T. J., and P. E. Kolattukudy. 1972. *Biochem. Biophys. Res. Commun.* **46**:16.

Wolff, I. A. 1966. *Science* **154**:1140.

Yamada, M., and P. K. Stumpf. 1964. *Biochem. Biophys. Res. Commun.* **14**:165.

Yang, S. F., and P. K. Stumpf. 1965. *Biochim. Biophys. Acta* **98**:27.

Yano, I., B. W. Nichols, L. J. Morris, and A. T. James. 1972. *Lipids* **7**:30.

Zimmerman, D. C. 1966. *Biochem. Biophys. Res. Commun.* **23**:398.

Zimmerman, D. C., and B. A. Vick. 1970. *Plant Physiol.* **46**:445.

SUBJECT INDEX

A

Acacia, nonprotein amino acids in, 97

Acer, nonprotein amino acids in, 97

Aesculus, nonprotein amino acids in, 97 111–113

Alanines, β-substituted, natural occurrence of, 100–101

Alkaloids
  biosynthesis of
    induction, 157–167
    precursors 151–157
    regulatory control, 141–178
  degradation of, 147–148
    induction, 172–173
  external factor effects on, 150
  feedback regulation of, 168–172
  genetic determinants for, 148–149
  physiological and evolutionary significance of, 174–175
  translocation and transport of, 144–147

Amino acids, nonprotein, 95–122

Aminoacyl-*t*RNA synthetase, in protein synthesis, 109–115

2-Amino-hex-4,5-dienoic acid, occurrence of, 111

2-Amino-6-hydroxy-4-methylhex-4-enoic acid, occurrence of, 111

2-Amino-4-methylhex-4-enoic acid, occurrence of, 111

L-4-Aminophenylalanine, natural occurrence of, 101

Anabasine, biosynthesis of, 151, 153–154

Anthocyanins, biosynthesis of, 3

Apigenin
  biosynthesis of, 26, 44
  glycosides of, 29

Apigeninidin, in *Sorghum*, 59

Apiin
  apiose from, 22, 24
  biosynthesis of, 23–24

Apiose, in parsley, 22

Apiosyltransferase, in flavonoid biosynthesis, 43–44

Artemisic acids, occurrence of, 219

Asclepic acid, natural occurrence of, 214, 224

Asparagines, in plants, 97

Aspergillus nidulans, tyrosinase from, 88

Astragalus, nonprotein amino acids in, 97

Atropa belladonna, alkaloid biosynthesis in, 149

Auricolic acid, occurrence of, 216

Azetidine-2-carboxylic acid, in plants, 98

B

Bacteria, permease systems in, 105–107

Baikiaea plurijuga, nonprotein amino acids in, 98

Baikiain, in plants, 98

Beet
  nonprotein amino acids in, 98
  prolyl-*t*RNA synthetase from, 116

Bolekic acid, occurrence of, 221

Bornesitol, in inositol biochemistry, 191, 193

C

Caesalpinia spp.; nonprotein amino acids in, 97, 99, 107

Caffeic acid, in *Sorghum*, 59

Calendic acid, occurrence of, 218

Canavanine, in plants, 97

N-Carbamoyl-DL-*p*-hydroxyphenylglycine, natural occurrence of, 101

D-3-Carboxy-4-hydroxyphenylglycine, natural occurrence of, 101

m-Carboxyphenylalanine, in plants, 97

L-*m*-Carboxyphenylalanine, natural occurrence of, 101

D-*m*-Carboxyphenylglycine, natural occurrence of, 101

m-Carboxytyrosine, biosynthesis of, 104

L-*m*-Carboxytyrosine, natural occurrence of, 101

Catalpic acid, occurrence of, 219

Chalcones, as chalcone-flavanone isomerase substrates, 38, 39

Chalcone-flavanone isomerase
  in flavonoid biosynthesis, 23, 29, 30, 34, 36–38
  isoenzymes of, 39
  properties of, 30

Chalcone-flavanone oxidase, in flavonoid biosynthesis, 29, 40–41

Chalcone-flavanone synthetase, in flavanoid biosynthesis, 29, 34–35

Chalcone synthetase, in flavonoid biosynthesis, 30

Chemotaxonomy, fatty acids in, 227–228

Chloramphenicol, effect on *Polyporus* enzymes, 92

Chlorogenic acid, biosynthesis of, 3

Chlorogenic acid oxidase, in phenolics biosynthesis, 63–64

Chorismate mutase, in alkaloid biosynthesis, 155

Chrysoeriol, biosynthesis of, 26

Chymotrypsin inhibitor I, isolation of, 124–125

Cinnamate 4-hydroxylase, in phenylpropanoid biosynthesis, 87, 89

Cinnamate pathway, of phenylalanine degradation, in fungi, 82

Cinnamic acid, biosynthesis of, 3

Cinnamic acid: CoA ligase, 23
  in flavanoid biosynthesis, 27

Cinnamic 4-hydroxylase, in phenolics biosynthesis, 15–16, 55, 63

Cinnamyl alcohols, polymerization to lignin, 49–50

Clavaria, phenylpropanoid biosynthesis in, 93

Claviceps, alkaloid biosynthesis in, 144, 145–146, 149, 151, 154–159, 174

Coniferyl alcohol, biosynthesis of, 48–49

Convallaria majalis, prolyl-*t*RNA synthetase from, 116

Convolvulinolic acid, natural occurrence of, 213

Coriolic acids, occurrence of, 219, 226, 235

Coronaric acid, occurrence of, 216

Cosmosiin, in flavonoid biosynthesis, 44

p-Coumarate : CoA ligase
  in flavonoid biosynthesis, 27, 32–33, 34
  in phenylpropanoid biosynthesis, 88

p-Coumarate hydroxylase, in hispidin synthesis, 87–88, 89

Coupeic acid, occurrence of, 219

Crepenynic acid, occurrence of, 218

Cutin, fatty acids in, 211, 222

Cyanidin, in *Sorghum*, 59

Cycloaldolase, in *myo*-inositol biosynthesis, 182

Cycloheximide
  effect on phenylalanine ammonia-lyase, 10, 11
  effect on *Polyporus* enzymes, 92

Cyclopenase
  in alkaloid biosynthesis, 166
  occurrence of, 144

Cyclopenin, biosynthesis of, 144, 145

Cyclopenol, biosynthesis of, 144, 145

Cyclopropane acids
  occurrence of, 223, 228
  in plant sulfolipid, 211

D

Damascine, biosynthesis of, 144

Datura spp., alkaloid biosynthesis in, 149

Dehydrocrepenynic acid, occurrence of, 219